DIRECTIONS FOR USING

Thacher's Calculating Instrument

Edwin Thacher M. Am. Soc.

KEUFFEL & ESSER CO.

NEW YORK

1907

PREFACE.

THE ordinary slide-rule, or Gunter's scale, invented many
years ago, for approximation or a rough check on ordinary calculation, has been found of great utility. It is, however, unfit-
ted for the more exact calculations of the engineer, architect, and
actuary, and does not satisfy all the requirements of scientific
investigation or of financial transactions.

The value of the slide-rule as a calculator depends upon the
accuracy of its results, upon the rapidity with which those results can be found, upon the variety of its useful applications,
and upon the small amount of mental and physical effort required in the solution of problems.

With any form of slide-rule the results are necessarily approximate, the degree of approximation depending upon the length of
scales, the care observed in their manufacture, and the material
used. With the ordinary construction and material it is not
practicable to make such length of scales as will give results beyond the third place, which latter is frequently in doubt, the
doubt being still further increased by the unequal shrinkage of
the material of which it is made, usually boxwood. The ordinary rule in use is 12 inches long, with radii of 11 and 5½ inches,
the divisions of which are cut by hand, copying from a machine-
divided plate. In the present instrument the radii are 60 and 30
feet, the divisions of which are printed directly from machine-di-
vided plates. These plates contain over 33,000 divisions, calculated to seven places of decimals from Babbage's tables by using
a common multiplier, every line being subjected to correction for
error of screw and temperature variation, so that possibly every line
centre is within .0001 inch of its true place. The plates contain
over 17,000 engraved figures. The work was executed by W. F.
Stanley, of London, upon a dividing-machine made expressly for

this rule, the diamond point used for cutting the divisions being moved by a screw of 50 threads per inch, reading by micrometer to .000001 inch. It was a work requiring much time, care, and skill. The framework of this rule is of metal throughout, therefore not injuriously affected by variations in temperature nor moisture, no time, pains, or expense having been spared to make the instrument in every way reliable ; the scales, being of great length, will give results correctly to four, and usually to five, places of figures, sufficient to satisfy nearly every requirement of the professional and business man. A knowledge of its use is readily acquired. The rules following are general, simple, and easily remembered. After once becoming familiar with the movements results are obtained with almost incredible rapidity ; ordinary calculations, and complicated formulæ involving roots and powers, being solved with nearly equal facility.

By the use of this instrument the drudgery of calculation is avoided, and the relief to the mind may be compared with the most improved mechanical appliances in overcoming the wear and tear of manual labor.

It must not be supposed that this instrument is of advantage to those only who have constant use for it. If a knowledge of its use is once acquired it will never be forgotten, and will always be available for instant application. If one has but little calculating to do this may not only be quickly done, but in cases of emergency, which may occur in almost any business, a power is held in reserve to be relied upon.

By use of this instrument results are more reliable than when brought out in the usual way ; mistakes are less liable to occur, and if they should occur the whole work can be quickly checked. The only liability to error is in the misreading of the figures and divisions upon the scales, which are essentially the same as the readings of the inches and parts of a carpenter's rule.

The useful applications of the instrument are as general as the fundamental rules of arithmetic. Whether considered as a novelty and used only as a means of recreation and amusement, or the object in view be information or practical utility in business, the time spent in its investigation will never be regretted. Professors and students of schools and colleges will find it the most valuable aid to mathematical study that can possibly be desired. As the reasoning powers are exercised in the correct statement of any question by indicating clearly the operations to be performed, the countless repetitions of already familiar arithmetical processes become extremely irksome, and often a waste of valuable time. A familiar knowledge of its use acquired at school will be found invaluable when applied to the multiplicity of business transactions in after-life.

The engineer will find ample opportunity for its use. The bridge engineer will find it useful in finding the moments and shears in plate-girders, the strains and sections in trusses, the bearing and shearing values of pins and rivets, the thickness of bearing plates, the length of tie-plates, and estimates of quantities, costs, etc.

The railway engineer will find it useful in the mensuration of lines, areas, and solids ; in the production of estimates of work to be done and of material to be furnished ; also in solving trigonometrical and other formulæ, in making and applying tables, etc.

The mechanical engineer will find it useful in problems involving any of the mechanical powers; in finding the pitch, number of teeth, and diameter of toothed wheels; in designing the parts of engines and other machinery, and of ascertaining their power, etc.

The hydraulic engineer will find it useful in ascertaining the force, velocity, and pressure of water; the power of water-wheels; in designs of pumping-engines, and the solution of most of the formulæ relating to hydrostatics and hydraulics.

The architect and actuary, not less than the engineer, will find it useful in their various calculations and estimates.

The scientist will find it useful in the conversion of thermometric scales; in finding the specific gravity of bodies; in problems relating to uniform or accelerated motion, force of wind, velocity of sound, vibrations of pendulums, centrifugal force, etc.

The navigator will find it useful in the solution of all formulæ pertaining to plane and spherical trigonometry.

The mechanic will find it useful in estimating the contents of round or square timber; in finding feet board measure, angles, diagonals, etc.

The business man will find it useful in problems of simple and compound interest, discount and fellowship, extending pay-rolls, pro-rating accounts, assessing taxes, gauging casks, estimating the weights of metals and other material, for the exchange of money, conversion of foreign and domestic weights and measures, etc.; and it will be found generally useful in miscellaneous application of multiplication, division, proportion, fractions, roots and powers, and as a check on ordinary calculations.

No attempt has been made to give numerical examples, only to such extent as will be sufficient to give a clear illustration of the rules and use of the instrument, with which no difficulty will be found in adapting it to any required special class of work.

Attention is called to the derivation and use of gauge-points, by means of which complex questions are often greatly simplified. Tables have been given to satisfy many requirements, but they may be extended indefinitely to suit special wants. A certain small portion only of the following directions, rules, and applications will be found to be a tax upon the memory. Rule I., which occupies but a few lines, covers the great majority of questions and should be thoroughly understood. The remaining rules are seldom used and can be committed to memory at leisure as required, and the same may be said of the tables of gauge-points for special applications.

The aim has been to make the subject clear and comprehensive, and it is hoped the instrument will be appreciated by our overtaxed computers in general.

CONTENTS.

PART I.

PART II.

GENERAL EXAMPLES FOR PRACTICE AND REFERENCE.

PART III.

SPECIAL APPLICATIONS.

8   *CONTENTS.*

DECIMAL EQUIVALENTS OF ENGLISH MEASURES.

EXPLANATION OF SIGNS.

+ Addition, called plus.	> Inequality, greater than.	$\sqrt{}$ Square root.
— Subtraction, " minus.	< " less than.	$\sqrt[3]{}$ Cube "
× Multiplication.	∴ Hence, or therefore.	a, b, c, d, Constants.
÷ Division.	: Is to.	x, y, z, u, Variables.
= Equality.	:: As.	

PART I.

SCALE READINGS.

THIS instrument offers a mechanical solution of problems due to the principle of logarithms that the product of two numbers corresponds to the sum of their logarithms, the quotient of two numbers to the difference of their logarithms, and the power of any number to the product of the logarithm of the number and the index of its power.

A logarithmic scale is one representing the logarithms of all numbers, or one on which the distance of any number from the commencement of the scale corresponds to the mantissa of its logarithm.

It may be supposed by some that, as the instrument is indebted to logarithms for its power, a knowledge of logarithms is essential to its intelligent manipulation ; but such is not the case. The numbers and divisions on the scales are not logarithms, but correspond in position to logarithms, and are read in the same way as any other decimal scale. With ordinary scales of equal parts the sliding of one scale upon another would give addition and subtraction. With logarithmic scales similar movements give multiplication and division. It may sometimes be interesting to follow out the manner in which different movements produce the desired result, but for all practical purposes the simple rules following will be found sufficient.

In the use of this instrument the first requirement is to be able to read correctly and rapidly the numbers and divisions on the scales, which are known as *A*, *B*, and *C*, and each line is thus marked. This presents no difficulty, but requires practice in order to become proficient.

The numbers 100, 101, 102, etc. . . . 1000 mark the principal divisions on the scales ; the marks to which they refer are known by their greater length, extending entirely through the space occupied by the numbers, and, on scales *A* and *B*, also by their greater width.

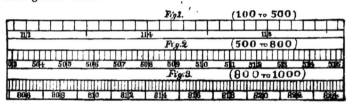

Fig.1.   (100 to 500)

Fig.2   (500 to 800)

Fig.3.   (800 to 1000)

Figs. 1, 2, and 3 show small portions of one of the scales in different parts of its length, and represent the only varieties of numbering and dividing to be found on them. All quantities on scales *A* and *B* between 100 and 500, and on scale *C* between 100 and 650, are shown like Fig. 1. All quantities on scales *A* and *B* between 500 and 800, and on scale *C* between 650 and 1000, are shown like Fig. 2, and all quantities on scales *A* and *B* between 800 and 1000 are shown like Fig. 3. Fig. 3 differs from Fig. 2 only in the omission of odd numbers, to avoid contracting or crowding.

All divisions are to be read decimally. In Fig. 1 each of the principal divisions is divided into ten parts, which represent all values of the fourth figure ; thus between 113 and 114 the first mark following 113 reads 113.1, the second mark 113.2, the fifth mark 113.5, the eighth mark 113.8, etc. The fifth or middle mark is longer than the others, to assist in reading. The values of the divisions on the scales are all arbitrary, the values set upon them depending upon the question considered ; thus 113 on the scales may mean 11300, 1130, 113, 11.3, 1.13, .113, .0113, etc., but the values of any intermediate divisions are governed by the principal divisions ; for instance, between 113 and 114, if 113 means 1130 the first mark following means 1131, the fourth mark 1134, etc. If 113 means 11.3 the first mark following means 11.31, the fifth mark 11.35, and so on for the fourth figure of all numbers between 1000 and 5000 on scales *A* and *B*, and between 1000 and 6500 on scale *C*.

In Figs. 2 and 3 each of the principal divisions is divided into five parts, each part representing two-tenths or the even values of the fourth figure, the odd values being midway between them; thus in Fig. 2, between 506 and 507, 506.1 is midway between 506 and the first mark following; 506.2 is the first mark, 506.3 is midway between the first and second mark, 506.4 is the second mark, 506.5 is midway between the second and third mark, 506.6 is the third mark, and so on for the fourth figure of all numbers between 5000 and 10000 on scales *A* and *B*, and between 6500 and 10000 on scale *C*.

Having found the fourth figure of any number, it remains to determine or approximate to the fifth. In Fig. 1, between 114 and 115, we have seen that if the first mark following 114 means 11410, the second mark means 11420. The space between these marks is conceived to be divided by the eye into ten equal parts, and the number of these parts, estimating from the left, corresponds to the value of the fifth figure; thus a point midway between these marks reads 11415, a point three-tenths of the space from the left reads 11413, a point seven-tenths from the left reads 11417, and so on for the fifth figure of all numbers between 10000 and 50000 on scales *A* and *B*, and between 10000 and 65000 on scale *C*. A little practice will enable the fifth figure to be found with as great exactness as would be possible were the points of division actually represented.

In Fig. 2, between 512 and 513, we have seen that if the first mark following 512 means 51220, the second mark means 51240. Any number between these must be estimated by the eye, the entire space representing twenty units; thus a point one-fourth space from the left reads 51225, a point at centre of space reads 51230, a point three-fourths space from the left reads 51235. One can readily judge by the eye as to one-fourth, one-half, or three-fourths of a space, also as to whether any required point represent a little more or a little less than any of these; and although it cannot be expected that the fifth figure can be found with absolute certainty in all cases, it can be closely approximated to.

The learner is earnestly advised to become thoroughly familiar with the reading of the scales, so that any number of five figures can be instantly located before proceeding further. This being done, but little difficulty will be found in applying the rules and directions following.

DESCRIPTION.

The instrument consists of a cylindrical slide provided with a knob at each end, and which admits of both rotary and longitudinal movement, within an open framework or envelope of equidistant bars of triangular section. The bars are connected ·to rings at their ends, which admit of rotation within standards attached to the base. The surface of the slide is exposed in the openings between the bars, the lower edges of which are in contact with it. The diameter and length of slide and the number of bars may be varied at pleasure, but in the present instrument the slide is four inches in diameter and has a divided length of eighteen inches, and the envelope contains twenty bars. Upon the slide are wrapped two complete logarithmic scales, one on each side of the centre. Each scale from 100 to 1000 is divided into forty parts of equal length; the number of parts equals the number of exposed sides of the bars, and the length of each is one-half the graduated length of the slide. These parts follow each other in regular order around the cylinder, both on the left and right, and those on the right follow those on the left in regular order. The figures and divisions which constitute any part on the right are repeated on the left one line in advance. All figures and divisions on the scales are made to face both ways, that they may be read from either edge of the bars.

Upon the lower lines of the bars, and in contact with the slide, are two other scales of the same length and arranged in the same manner as those on the slide, there being a complete scale both on the left and right of the centre. When the commencement of the scales on the slide and envelope are in contact all divisions on the one are opposite corresponding divisions of the other. Upon the upper lines of the bars, and not in contact with the slide, is a scale of roots. This scale is double the length of the others, and is divided into double the number of equal parts. It occupies the entire upper part of the bars, both on the left and right of the dividing centre line. These parts are laid off in regular order, and in such manner that any number on the lower line of bar is the square of the number opposite to it on the upper line, and any number on the upper line of bar is the square root of the number opposite to it on the lower line. For numbers having 1, 3, 5, etc., or an odd number of places, their roots are found on the left, and for numbers having 2, 4, 6, etc., or an even number of places, their roots are found on the right. By the rotary movement of the slide any line on it may be brought opposite to any line on the envelope, and by the longitudinal movement any graduations or subdivisions of these lines may be brought opposite to, or in contact with, each other. One-fortieth of a revolution is equivalent to nine inches of longitudinal movement.

When the slide is set in position for use the divisions on it appear of about equal length on both edges of the bars, and two lines of figures marking their values appear in each opening between the bars, two similar lines also being covered by each bar. The divisions on the upper lines are transferred to the slide by means of a pointer fitting over the bars. This pointer is also convenient for retaining the position of any division on either line while the slide is being revolved into the required position.

Near the commencement of each scale on the envelope and slide is a heavy black mark, designed to readily catch the eye during the rapid movement of the parts. The former will be found to expedite multiplication, and the latter division.

The scales for this instrument have a length forty times as great as for an ordinary slide-rule of the same length, and more than sixty-five times as great as for the ordinary slide-rule in use.

LOCATION AND RELATION OF SCALES A, B, AND C.

The scales on the slide will be known as A, those on the lower lines of the bars as B, and the one on the upper lines of the bars as C. If we represent values by the scales on which they are found—first values by A, B, and C respectively, and other values by A_1, B_1, and C_1 respectively—any value on one scale bears to the value opposite on another the following relation, the slide being in any position :

First. When the slide is direct, the numbers reading in the same direction as on the bars,

$$B : A :: B_1 : A_1 \therefore \frac{AB_1}{B} = A_1, \quad B : A :: C^2 : A_1 \therefore \frac{AC^2}{B} = A_1$$

$$C^2 : A :: B : A_1 \therefore \frac{AB}{C^2} = A_1,$$

$$C^2 : A :: C_1^2 : A_1 \therefore \frac{AC_1^2}{C^2} = A_1, B = C^2$$

$$\sqrt{A} : \sqrt{B} :: \sqrt{A_1} : C \therefore \sqrt{\frac{BA_1}{A}} = C,$$

$$\sqrt{A} : C :: \sqrt{A_1} : C_1 \therefore \sqrt{\frac{C^2A_1}{A}} = C_1, C = \sqrt{B}$$

Second. When the slide is reversed, the numbers reading from right to left.

. If 100 on *A* is set opposite 100 on *B*, then any number on *A* is the reciprocal of the number opposite on *B*, and vice versa.

Also, any number on *A* is the reciprocal of the square of the number opposite on *C*.

If the slide is in any position, any value on one scale bears to the value opposite on another the following relation :

$$\text{Rec. } B : A :: \text{Rec. } B_1 : A_1 \therefore \frac{AB}{B_1} = A_1,$$

$$\text{Rec. } C^2 : A :: \text{Rec. } B : A_1 \therefore \frac{AC^2}{B} = A_1$$

$$\text{Rec. } B : A :: \text{Rec. } C^2 : A_1 \therefore \frac{AB}{C^2} = A_1,$$

$$\text{Rec. } C^2 : A :: \text{Rec. } C_1^2 : A_1 \therefore \frac{AC^2}{C_1^2} = A_1$$

$$\sqrt{\text{Rec. } A} : \sqrt{B} :: \sqrt{\text{Rec. } A_1} : C \therefore \sqrt{\frac{BA}{A_1}} = C,$$

$$\sqrt{\text{Rec. } A} : C :: \sqrt{\text{Rec. } A_1} : C_1 \therefore \sqrt{\frac{C^2A}{A_1}} = C_1$$

The above relation of scales is a key to the rules hereafter given. It is seen that the solution of any question in multiplication, division, proportion, powers, and roots is but an application of the rule of three—that is to say, the answer is the fourth term of a proportion, and the statement contains the remaining terms. If $b : a :: x :$ answer, then $\dfrac{ax}{b} =$ answer, in which the denominator is the first term of the proportion, and the numerator contains the second and third. These terms may have any value, but in order that the rules may be simple and general the form of expression must be considered to remain unchanged. If $b = 1$, we have $\dfrac{ax}{1}$ or multiplication; if $a = 1$, we have $\dfrac{1 . x}{b}$ or division ; and if b, a, and x have a value different from 1, we have $\dfrac{ax}{b}$ or simple proportion. And the same applies to formulæ for powers and roots, and for direct or reversed slide.

RULES AND DIRECTIONS FOR OPERATING THE INSTRUMENT.

The formulæ

$$\left.\begin{array}{c} \dfrac{ax}{b},\ \dfrac{ax^2}{b},\ \dfrac{ax}{b^2},\ \text{and}\ \dfrac{ax^2}{b^2} \\[2mm] \dfrac{ba}{x},\ \dfrac{b^2a}{x},\ \dfrac{ba}{x^2},\ \text{and}\ \dfrac{b^2a}{x^2} \end{array}\right\} \begin{array}{l} \text{when the slide is direct} \\[1mm] (1) \\[1mm] \text{when the slide is reversed} \end{array}$$

are readily solved. a and b are used in setting. They may have any value, but the slide must be set as often as their value is changed. x is not used in setting, and may have any number of values without resetting. This feature is of great value in all pro-rata questions, taxes, fellowship, exchange, strains, sections, etc.—in fact, a large percentage of all calculations.

General Rule for the Solution of all Formulæ (1).

Rule I. Opposite b on $\left\{\begin{array}{l} B \text{ if first power} \\ C \text{ if second power} \end{array}\right\}$ set a on A, then opposite x on $\left\{\begin{array}{l} B \text{ if first power} \\ C \text{ if second power} \end{array}\right\}$ find the answer on A.

This rule applies whether the slide is direct with the variable in the numerator, or the slide is reversed with the variable in the denominator, and is sufficient to work all ordinary questions. It is of the first importance, therefore, that it be thoroughly impressed upon the memory.

In the statement of any question in multiplication, division, proportion, or powers it will be observed that the numerator contains two terms and the denominator one. It will be observed, further, that the slide, *A*, represents one factor of the numerator and the answer, and that the bar, *B* or *C*, represents the denominator and the second factor of the numerator.

If the slide is direct, as is usually the case, one factor of the numerator on the slide, *A*, is set opposite the denominator on the bar, *B* or *C*, then opposite the other factor of the numerator on the bar, *B* or *C*, the answer is found on the slide.

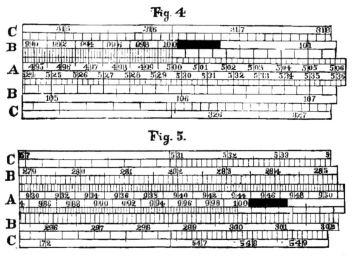

Fig. 4

Fig. 5.

If the slide is reversed the second factor of the numerator and the denominator are reversed, or one factor of the numerator on the slide, *A*, is set opposite the other factor of the numerator on the bar, *B* or *C*, then opposite the denominator on the bar, *B* or *C*, the answer is found on the slide. It will be remembered that both in setting and reading the lower line of bar, *B*, is used for first powers, and the upper line of bar, *C*, is used for second powers.

As the first three figures are printed on the rule, it will be found advantageous in setting or reading to temporarily disregard the decimal point and to group the first three figures. If, for instance, it is desired to set 46.94 opposite 2.1873, look for 218.73, opposite to which set 469.4.

Figs. 4 and 5 represent small portions of the scales, *A* being on the convex surface of the slide, *B* and *C* on the inclined sides of the bars, all shown developed.

A few examples, such as can be illustrated by these figures, will be given in application of Rule 1 (see Formulæ 1):

Multiplication, $\dfrac{ax}{b} = \dfrac{ax}{1}$ Ex. $5 \times 106 = \dfrac{5 \times 106}{1} = 530$ (see Fig. 4). Opposite 1 on *B* is set 5 on *A*, then opposite 106 on *B* read 530 on *A*. Also, opposite any other number on *B* read 5 times that number on *A*.

Ex. $5 \times 107 = 535$; $5 \times 105.5 = 527.5$; $5 \times 99.6 = 498$, etc.

Division, $\dfrac{ax}{b} = \dfrac{1 \, x}{b}$ Ex. $\dfrac{282}{3} = \dfrac{1 \times 282}{3} = 94$ (see Fig. 5). Opposite 3 on *B* is set 1 on *A*, then opposite 282 on *B* read 94 on *A*. Also, opposite any other number on *B* read one-third of that number on *A*.

Ex. $\dfrac{285}{3} = 95$, $\dfrac{299}{3} = 99.667$, $\dfrac{302}{3} = 100.67$, etc.

Proportion, $\dfrac{ax}{b}$ Ex. $\dfrac{49.9 \times 106.9}{99.8} = 53.45$ (see Fig. 4). Opposite 99.8 on *B* is set 49.9 on *A*, then opposite 106.9 on *B* read 53.45 on *A*. Also, opposite any other number on *B* read the ratio to which the instrument is set multiplied by that number on *A*.

Ex. $\dfrac{49.9 \times 101.1}{99.8} = 50.55$.

Powers, $\dfrac{ax^2}{b}$ Ex. $\dfrac{50.3 \times 31.5^2}{100.6}$ $= 496.12$ (see Fig. 4).   Opposite 100.6 on B is set 50.3 on A, then opposite 31.5 on C read
496.12 on A, and opposite any other number on C read the ratio
to which the instrument is set multiplied by the square of that
number on A.

Take $\dfrac{ax}{b^2}$ Ex. $\dfrac{53.5 \times 105.6}{32.71^2} = 5.28$ (see Fig. 4).   Opposite
32.71 on C is set 53.5 on A, then opposite 105.6 on B read 5.28
on A.

Take $\dfrac{ax^2}{b^2}$ Ex. $\dfrac{9.47 \times 17.26^2}{5.33^2} = 99.306$ (see Fig. 5).   Opposite 5.33 on C is set 9.47 on A, then opposite 17.26 on C read
99.306 on A.

In solving any of the formulæ (1) the following directions are
to be observed : Turn the instrument in the standards until b ap-
pears in convenient position on one of the bars.

On $\begin{Bmatrix} B \text{ if first power} \\ C \text{ if second power} \end{Bmatrix}$ hold the envelope in about this
position, and turn the slide until a is seen facing the line of bar
on which b was found; then move the slide backward or forward
until a is brought opposite to, or in contact with, b (use the pointer, if necessary).   The instrument is now set, and need not be
changed so long as a and b remain the same.   Leave the slide
in this position, and turn the envelope in the standards until x is
seen on the bar $\begin{Bmatrix} B \text{ if first power} \\ C \text{ if second power} \end{Bmatrix}$ opposite to which on A
the answer will be found.

The setting and reading can be accomplished with great rapidity after once becoming familiar with the movements.   The
rotary movement need never exceed half a revolution, nor the
longitudinal movement more than one-half the length of the
slide.

In setting the instrument for multiplication and division, although the commencement of scales (100) on the envelope and
slide may be used at the left, centre, or right, at pleasure, the

centre is preferable, as it requires a minimum of longitudinal movement.

Position of Decimal Point.

This can usually be found by mere inspection, or by a previous knowledge of the approximate value of any result ; but when not readily determined by these means it may always be found by observing the following directions :

First. In the statement of any question it must be borne in mind that the numerator contains two terms and the denominator one.

Secondly. Quantities of the first degree are taken plus the number of places of whole numbers, and minus the number of ciphers in decimals; thus $214.6 = +3$, $21.46 = +2$, $2.146 = +1$, $.2146 = 0$, $.02146 = -1$, $.002146 = -2$, etc. Quantities of the second degree found on the left half of lines C, or between 100 and 316.22, are taken plus double the number of places of whole numbers, less 1, and minus double the number of ciphers in decimals, less 1 ; thus $217.0^2 = +5$, $21.7^2 = +3$, $2.17^2 = +1$, $.217^2 = -1$, $.0217^2 = -3$, $.00217^2 = -5$, etc.

Quantities of the second degree found on the right half of lines C, or between 316.22 and 1000, are taken plus double the number of places of whole numbers, and minus double the number of ciphers in decimals; thus $514.0^2 = +6$, $51.4^2 = +4$, $5.14^2 = +2$, $.514^2 = 0$, $.0514^2 = -2$, $.00514^2 = -4$, etc.

The following is a general rule for finding the position of decimal point in all examples of the first or second degree :

Rule II. First, when the slide is direct—

If the number on A (regardless of decimal point), both in setting and reading, is either greater or less than the number opposite on B, the number of places in the result is the algebraic difference between the sum of the number of places in the numerator and in the denominator.

If the number on A is greater in setting and less in reading than the number opposite on B, the number of places in the result is the algebraic difference, as above, plus one.

If the number on A is less in setting and greater in reading than the number opposite on *B*, the number of places in the result is the algebraic difference, as above, minus 1.

The above rule may be expressed as follows :

Setting.		Reading.	Result.		
$A > B$	and	$A > B$	Algebraic difference.		
$A < B$	"	$A < B$	"	"	
$A > B$	"	$A < B$	"	"	+1.
$A < B$	"	$A > B$	"	"	−1.

Second, when the slide is reversed—

The number on *A* in setting is compared with the number on *B* in reading, and the number on *A* in reading with the number on *B* in setting. Otherwise the rule remains the same.

Examples in application of Rule II. :

$$\frac{6.12 \times 189.0}{20.6} = 56.15 \quad \begin{cases} \text{Alg. diff.} = (1+3)-2 = +2 \\ 612 > 206 \text{ and } 561 > 189 \\ \therefore \text{ Result} = \text{Alg. diff.} = +2 \text{ places.} \end{cases}$$

$$\frac{151.0 \times .0065}{.052} = 18.875 \quad \begin{cases} \text{Alg. diff.} = (3-2)+1 = +2 \\ (--1 = +1) \\ 151 < 520 \text{ and } 188 < 650 \\ \therefore \text{ Result} = \text{Alg. diff.} = +2 \text{ places.} \end{cases}$$

$$\frac{41.2 \times .0733}{26.4} = 0.1144 \quad \begin{cases} \text{Alg. diff.} = (2-1)-2 = -1 \\ 412 > 264 \text{ and } 114 < 733 \\ \therefore \text{ Result} = \text{Alg. diff.} + 1 = 0. \end{cases}$$

$$\frac{.031 \times 13.2}{.0046} = 88.956 \quad \begin{cases} \text{Alg. diff.} = (-1+2)+2 = +3 \\ 31 < 46 \text{ and } 889 > 132 \\ \therefore \text{ Result} = \text{Alg. diff.} - 1 = +2 \text{ places.} \end{cases}$$

$$\frac{16.0 \times 141.0}{1} = 2256.0 \quad \begin{cases} \text{Alg. diff.} = (2+3)-1 = 4 \\ 16 > 10 \text{ and } 225 > 141 \\ \therefore \text{ Result} = \text{Alg. diff.} = +4 \text{ places.} \end{cases}$$

The above examples contain only first powers, and, as the numbers on *A* and *B* mentioned in Rule II. appear in the statements, the rule can be applied to ordinary arithmetical workings without special reference to the instrument ; but in the examples which follow, and contain the second power, line C is used in the statement, and the number on line B immediately below it on the bar is used in the rule.   In this case the instrument must be consulted, unless the number on B, which is the square of the number on C in the statement, can be found mentally, with sufficient approximation to satisfy the requirement of the rule.

The examples containing second powers previously given under Rule I. will be repeated here, in order that Figs. 4 and 5 may be used in illustration.

$$\frac{50.3 \times 31.5^2}{100.6} = 496.12 \left\{ \begin{array}{l} \text{(See Fig. 4.) Alg. diff.} = (2+3)-3 = +2 \\ 503 > 100 \text{ and } 496 < 992 \\ \therefore \text{ Result} = \text{Alg. diff.} + 1 = +3 \text{ places.} \end{array} \right.$$

Note.—In the above, $3^2 = 9$, obtained mentally, could be substituted for 992, and be near enough to satisfy the requirements of the rule.

$$\frac{53.5 \times 105.6}{32.71^2} = 5.28 \left\{ \begin{array}{l} \text{(See Fig. 4.) Alg. diff.} = (2+3)-4 = +1 \\ 535 > 107 \text{ and } 528 > 105 \\ \therefore \text{ Result} = \text{Alg. diff.} = +1 \text{ place.} \end{array} \right.$$

$$\frac{9.47 \times 17.26^2}{5.33^2} = 99.306 \left\{ \begin{array}{l} \text{(See Fig. 5.) Alg. diff.} = (1+3)-2 = +2 \\ 947 > 284 \text{ and } 993 > 298 \\ \therefore \text{ Result} = \text{Alg. diff.} = +2 \text{ places.} \end{array} \right.$$

The formulæ

$$\left. \begin{array}{l} \sqrt{\dfrac{bx}{a}} \text{ and } \sqrt{\dfrac{b^2x}{a}} \quad \text{when the slide is direct} \\[1.5em] \sqrt{\dfrac{ba}{x}} \text{ and } \sqrt{\dfrac{b^2a}{x}} \quad \text{when the slide is reversed} \end{array} \right\} (2)$$

are solved by the following general rule :

Rule III. Opposite b on $\left\{ \begin{array}{l} B \text{ if first power} \\ C \text{ if second power} \end{array} \right\}$ set a on A, then opposite x on A read the answer on C.

In applying Rule III. it must be remembered that radicals have two roots, one for an odd and one for an even number of places under the $\sqrt{}$. If odd the root is found on the left, and if even on the right of the centre of lines *C*. Unless, therefore, the number of places is evident by inspection, or, as is generally the case, the answer is known sufficiently near to avoid mistake, it will be better to first find the value under the $\sqrt{}$, and then find the root by a second operation.

Ex. $\sqrt{\dfrac{9.98 \times 5.34}{4.99}} = 3.268$ (see Fig. 4). Opposite 9.98 on *B* is set 4.99 on *A*, then opposite 5.34 on *A* read 3.268 on *C*.

Ex. $\sqrt{\dfrac{5.33^2 \times 10.01}{9.47}} = 5.48$ (see Fig. 5). Opposite 5.33 on *C* is set 9.47 on *A*, then opposite 10.01 on *A* read 5.48 on *C*.

Cube Roots.

Rule IV. Reverse the slide. Set the number on *A* opposite 1 on *B* or *C*, then look for the number on *A* opposite the same number on *C*. This number is the cube root sought.

As there are three points on the lines that satisfy this condition, it is necessary to know approximately the first figure, or the wrong root is liable to be taken.

If the first figure in the result is 3 or less, set to *C* on the left, and if more than 3 to *C* on the right.

The application of the above rule requires a little patience; still, after becoming familiar with the instrument, the time required for finding the correct root of any number will not usually exceed one minute.

Ex. $\sqrt[3]{8.41} = 2.0336$, $\sqrt[3]{84.1} = 4.3815$, $\sqrt[3]{841.0} = 9.4391$.

Reverse the slide. Set 841 on *A* opposite 1 on *B*, then opposite 2.0336 on *C* find 2.0336 on *A*, opposite 4.3815 on *C* find 4.3815 on *A*, and opposite 9.4391 on *C* find 9.4391 on *A*.

Factors Composed of More than One Quantity.

In Formulæ (1) each factor of the numerator consists of one

quantity, but one of them may be composed of two or more quantities of the first or second degree connected by the signs + or —, as $\dfrac{a(x+y+z^2+u^2)}{b}$, $\dfrac{a(x^2+y+z^2)}{b^2}$, etc.

Rule· I. applies. It is only necessary to take a reading opposite each part of the second factor and add the results.

Ex. $\dfrac{2.75(47.3+61.8+9.71^2+0.67^2)}{6.41^2} = 31.658+41.363+63{,}101 - .300 = 136.422.$

Opposite 641 on *C* set 275 on *A*, then opposite 473 on *B* read 31.658 on *A*. Opposite 618 on *B* read 41.363 on *A*, opposite 971 on *C* read 63.101 on *A*, and opposite 670 on *C* read .30045 on *A*.

To Multiply Three Numbers together by One Operation.

Three numbers may be multiplied together by using the reciprocal of one of the numbers, for to multiply by a number is the same as to divide by its reciprocal :

$$bax = \dfrac{ax}{Rec.\ b} \qquad \text{Ex. } 4\times7.5\times61.3 = \dfrac{7.5\times61.3}{.25} = 1839.$$

Opposite the reciprocal of $4\ (\frac{1}{4}=.25)$ on *B* set 7.5 on *A*, then opposite 61.3 on *B* find 1839 on *A*.

For additional examples and table of reciprocals see Weights of Wrought Iron and Steel.

Formulæ Requiring More than One Setting.

Any of the formulæ heretofore given requires but one setting of the instrument, and as many readings as the variable has values, but formulæ of a more complicated nature may be factored and worked by two or more settings ; thus $\dfrac{ax^2}{b^2c} = \dfrac{ax^2}{b^2} \times \dfrac{x}{c}$ requires two settings, $\dfrac{a \times b \times c \times d}{e \times f \times g} = \dfrac{a \times b}{e} \times \dfrac{c}{f} \times \dfrac{d}{g}$ requires three settings, etc.

Ex. $\dfrac{16.3 \times 9.72 \times 81.6 \times 2.17^2}{7.83 \times 1.94 \times 63.1} = 63.512.$

Opposite 783 on *B* set 163 on *A*, then opposite 972 on *B* read 20.234 on *A*. Opposite 194 on *B* set 20.234 on *A*, then opposite 816 on *B* read 351.08 on *A*. Opposite 631 on *B* set 351.08 on *A*, then opposite 217 on *C* read 63.512 on *A*.

Rule II. Alg. diff. $= (2 + 1 + 2 + 1) - (1 + 1 + 2) = +2$, $< <$, $> >$, $> >$. ∴ Result = Alg. diff. $= +2$ places.

PART II.

GENERAL EXAMPLES FOR PRACTICE AND REFE-
RENCE.

Slide Direct.

$\frac{ax}{b}$ Ex. $\dfrac{16.0 \times .013}{41.0} = .005073.$

Rule I. Opposite 410 on B set 160 on A, then opposite 130 on B read 5073 on A.

Rule II. Alg. diff. $= (2-1) - 2 = -1$, $160 < 410$ and $507 > 130$, $\therefore$ Result $=$ Alg. diff. $-1 = -2$ places.

ax Ex. $4.37 \times 9.68 = \dfrac{4.37 \times 9.68}{1} = 42.301.$

Rule I. Opposite 1 on B set 437 on A, then opposite 968 on B read 4.2301 on A.

Rule II. Alg. diff. $= (1+1) - 1 = +1$, $437 > 1$ and $423 < 968$, $\therefore$ Result $=$ Alg. diff. $+1 = +2$ places.

$\frac{x}{b}$.   Ex. $\dfrac{581\ 3}{27.6} = \dfrac{1 \times 581.3}{27.6} = 21.061.$

Rule I. Opposite 276 on B set 1 on A, then opposite 581.3 on B read 21061 on A.

Rule II. Alg. diff. $= (1+3) - 2 = +2$, $100 < 276$ and $210 < 581$, $\therefore$ Result $=$ Alg. diff. $= +2$ places.

$\frac{ax'}{b}$ Ex. $\dfrac{224 \times 7.641'}{91.8} = 142.46.$

Rule I. Opposite 918 on B set 224 on A, then opposite 7641 on C read 14246 on A.

Rule II. Alg. diff.$=(3+2)-2=+3$, $224<918$ and $142<584$, ∴ Result $=$ Alg. diff. $=+3$ places.

ax^2 *Ex.* $74.1\times6.83^2=\dfrac{74.1\times6.83^2}{1}=3456.7.$

Rule I. Opposite 100 on *B* set 741 on *A*, then opposite 683 on *C* read 34567 on *A*.

Rule II. Alg. diff.$=(2+2)-1=+3$, $741>100$ and $345<466$, ∴ Result $=$ Alg. diff.$+1=4$ places.

$\dfrac{x^2}{b}$ *Ex.* $\dfrac{21.63^2}{14.1}=\dfrac{1\times21.63^2}{14.1}=33.181.$

Rule I. Opposite 141 on *B* set 100 on *A*, then opposite 2163 on *C* read 33181 on *A*.

Rule II. Alg. diff.$=(1+3)-2=+2$, $100<141$ and $331<468$, ∴ Result $=$ Alg. diff. $=+2$ places.

$\dfrac{a^2}{b}$ *Ex.* $\dfrac{23.0^2}{7}=\dfrac{23.0\times23.0^2}{7}=1738.1.$

Rule I. Opposite 700 on *B* set 230 on *A*, then opposite 230 on *C* read 17381 on *A*.

Rule II. Alg. diff.$=(2+3)-1=+4$, $230<700$ and $173<529$, ∴ Result $=$ Alg. diff. $=+4$ places.

a^3 *Ex.* $13.4^3=\dfrac{13.4\times13.4^2}{1}=2406.1.$

Rule I. Opposite 100 on *B* set 134 on *A*, then opposite 134 on *C* read 24061 on *A*.

Rule II. Alg. diff.$=(2+3)-1=+4$, $134>100$ and $240>179$, ∴ Result $=$ Alg. diff.$=+4$ places.

$\dfrac{ax}{b^2}$ *Ex.* $\dfrac{14.6\times.0063}{0.71^2}=0.18246.$

Rule I. Opposite 710 on *C* set 146 on *A*, then opposite 630 on *B* read 18246 on *A*.

Rule II. Alg. diff. $= (2-2) - 0 = 0$, $146 < 504$ and $182 < 630$, $\therefore$ Result $=$ Alg. diff. $= 0$ places.

$\dfrac{x}{b^2}$ Ex. $\dfrac{0.1463}{.00471^2} = \dfrac{1 \times 0.1463}{.00471^2} = 6594.8$.

Rule I. Opposite 471 on C set 100 on A, then opposite 1463 on B read 65948 on A.

Rule II. Alg. diff. $= (1+0) + 4 = +5$, $100 < 221$ and $658 > 146$, $\therefore$ Result $=$ Alg. diff. $- 1 = +4$ places.

$\dfrac{ax^2}{b^2}$ Ex. $\dfrac{6.46 \times 18.63^2}{218.1^2} = .047134$.

Rule I. Opposite 2181 on C set 646 on A, then opposite 1863 on C read 47134 on A.

Rule II. Alg. diff. $= (1+3) - 5 = -1$, $646 > 476$ and $471 > 347$, $\therefore$ Result $=$ Alg. diff. $= -1$ place.

$\dfrac{x^2}{b^2}$ Ex. $\dfrac{.0046^2}{.0874^2} = \dfrac{1 \times .0046^2}{.0874^2} = .00277$.

Rule I. Opposite 874 on C set 100 on A, then opposite 460 on C read 27700 on A.

Rule II. Alg. diff. $= (1-4) + 2 = -1$, $100 < 764$ and $277 > 212$, $\therefore$ Result $=$ Alg. diff. $- 1 = -2$ places.

$\dfrac{a^2}{b^2}$ Ex. $\dfrac{71.4^2}{86.3^2} = \dfrac{71.4 \times 71.4^2}{86.3^2} = 48.87$.

Rule I. Opposite 863 on C set 714 on A, then opposite 714 on C read 4887 on A.

Rule II. Alg. diff. $= (2+4) - 4 = +2$, $714 < 745$ and $488 < 510$, $\therefore$ Result $=$ Alg. diff. $= +2$ places.

$\sqrt{\dfrac{bx}{a}}$ Ex. $\sqrt{\dfrac{7.5 \times 19.4}{6.75}} = 4.6428$.

Rule III. Opposite 750 on B set 675 on A, then opposite 194 on A read 4.6428 on C.

$\sqrt{bx}$ Ex. $\sqrt{30.5 \times 7.63} = \sqrt{\dfrac{30.5 \times 7.63}{1}} = 15.255 = $ mean proportional.

Rule III. Opposite 305 on *B* set 100 on *A*, then opposite 763 on *A* read 15.255 on *C*.

$\sqrt{\dfrac{x}{a}}$ Ex. $\sqrt{\dfrac{425.1}{36.8}} = \sqrt{\dfrac{1 \times 425.1}{36.8}} = 3.3987.$

Rule III. Opposite 100 on *B* set 368 on *A*, then opposite 4251 on *A* read 3.3987 on *C*.

$\sqrt{\dfrac{b^2 x}{a}}$ Ex. $\sqrt{\dfrac{53.5^2 \times 8.61}{31.5}} = 27.97.$

Rule III. Opposite 535 on *C* set 315 on *A*, then opposite 861 on *A* read 27.97 on *C*.

$\sqrt{\dfrac{b^2}{a}}$ Ex. $\sqrt{\dfrac{225.6^2}{16.0}} = \sqrt{\dfrac{1 \times 225.6^2}{16.0}} = 56.4.$

Rule III. Opposite 225.6 on *C* set 160 on *A*, then opposite 100 on *A* read 56.4 on *C*.

$\sqrt{\dfrac{b^3}{a}}$ Ex. $\sqrt{\dfrac{18.3^3}{6.5}} = \sqrt{\dfrac{18.3 \times 18.3^2}{6.5}} = 30.705.$

Rule III. Opposite 183 on *C* set 650 on *A*, then opposite 183 on *A* read 30.705 on *C*.

$\sqrt{b^3}$ Ex. $\sqrt{24.3^3} = \sqrt{\dfrac{24.3 \times 24.3^2}{1}} = 119.79.$

Rule III. Opposite 243 on *C* set 100 on *A*, then opposite 243 on *A* read 119.79 on *C*.

It will be observed in the above examples under Rule III. that opposite the numerator on the bar is set the denominator on the slide, and that the answer is found on the bar, being the reverse of Rule I.

Slide Reversed.

$\dfrac{ba}{x}$ Ex. $\dfrac{12.1 \times 24.7}{37.1} = 8.055$.

Rule I. Opposite 121 on *B* set 247 on *A*, then opposite 371 on *B* read 8055 on *A*.

Rule II. Alg. diff. $= (2+2) - 2 = +2$, $247 < 371$ and $805 > 121$, ∴ Result $=$ Alg. diff. $-1 = +1$ place.

$\dfrac{b^2 a}{x}$ Ex. $\dfrac{.063^2 \times 13.4}{.0027} = 19.698$.

Rule I. Opposite 630 on *C* set 134 on *A*, then opposite 270 on *B* read 19698 on *A*.

Rule II. Alg. diff. $= (-2 + 2) + 2 = +2$, $134 < 270$ and $196 < 384$, ∴ Result $=$ Alg. diff. $= +2$ places.

$\dfrac{ba}{x^2}$ Ex. $\dfrac{28.6 \times 9.4}{8.7^2} = 3.5518$.

Rule I. Opposite 286 on *B* set 940 on *A*, then opposite 870 on *C* read 35518 on *A*.

Rule II. Alg. diff. $= (2+1) - 2 = +1$, $940 > 757$ and $355 > 286$, ∴ Result $=$ Alg. diff. $= +1$ place.

$\dfrac{b^2 a}{x^2}$ Ex. $\dfrac{3.81^2 \times 50.5}{7.4^2} = 13.387$.

Rule I. Opposite 381 on *C* set 505 on *A*, then opposite 740 on *C* read 13387 on *A*.

Rule II. Alg. diff. $= (2+2) - 2 = +2$, $505 < 548$ and $133 < 145$, ∴ Result $=$ Alg. diff. $= +2$ places.

$\sqrt{\dfrac{ba}{x}}$ Ex. $\sqrt{\dfrac{9.0 \times 8.0}{2}} = 6.0$.

Rule III. Opposite 900 on *B* set 800 on *A*, then opposite 200 on *A* read 6.0 on *C*.

$$\sqrt{\frac{b^2a}{x}} \quad \text{Ex.} \quad \sqrt{\frac{28.0^2 \times 9.81}{4.5}} = 41.341.$$

Rule III. Opposite 280 on C set 981 on A, then opposite 450 on A read 41.341 on C.

By making $a = 1$, or $a = b$, we have the following varieties of the above:

$$\frac{b}{x}, \frac{b^2}{x}, \frac{a^2}{x}, \frac{b}{x^2}, \frac{b^2}{x^2}, \frac{b^3}{x^2}, \sqrt{\frac{b^3}{x}}, \sqrt{\frac{b}{x}}, \sqrt{\frac{b^2}{x}}$$

There is no occasion to reverse the slide, except when a and b are constant and x has more than one value.

If the slide is direct and the denominator variable, it will be necessary to set the instrument for each value of x.

SQUARES, SQUARE ROOTS, AND DIAGONALS.

In questions of this nature the lines B and C only are used. Any number on B is the square of the number opposite on C, and any number on C is the square root of the number opposite on B. It must be remembered that roots of numbers having 1, 3, 5, etc., or an odd number of places, are found on the left half, and that roots of numbers having 2, 4, 6, etc., or an even number of places, are found on the right half, of lines C; also, that the square of any number less than 31622, disregarding decimal point, is found on the left half, and the square of any number greater than 31622 is found on the right half, of lines B.

Examples—Squares.

Left—less than 31622.		*Right—greater than 31622.*	
Lines C.	Lines B.	Lines C.	Lines B.
$11.4^2 = 129.96$		$3.21^2 = 10.304$	
$17.8^2 = 316.84$		$41.2^2 = 1697.4$	
$2.47^2 = 6.1009$		$5.97^2 = 35.641$	
$30.9^2 = 954.8$		$70.8^2 = 5012.6$	

Examples—Square Roots.

Left—odd number of places.   *Right—even number of places.*

Lines B.   Lines C.   Lines B.   Lines C.

$\sqrt{2.47} = 1.5716$ $\sqrt{24.7} = 4.9699$

$\sqrt{381.0} = 19.519$ $\sqrt{3810} = 61.725$

$\sqrt{12584} = 112.11$ $\sqrt{125840} = 354.74$

Examples—Diagonals.

$\sqrt{17.43^2 + 26.4^2} = 31.634.$

Opposite 17.43 on C find 303.80 on B.

     " 26.4 " C " 696.96 " B.

Sum $= 1000.76$

Opposite 1000.76 on B, right, find 31.634 on C.

$\sqrt{25.34^2 + 71.63^2} = 75.98.$

HIGHER POWERS AND ROOTS.

$a^3 = a \times a^2$. Apply Rule I.

Ex. $7.43^3 = \dfrac{7.43 \times 7.43^2}{1} = 410.17.$

$a^4 = a^2 \times a^2$. Take square twice by use of lines B and C.

Ex. 6.8^4 : $6.8^2 = 46.24$, and $46.24^2 = 2138.1.$

$a^5 = a \times a^2 \times a^2$. Apply Rule I. twice.

Ex. 2.4^5 : $\dfrac{2.4 \times 2.4^2}{1} = 13.824$, and $\dfrac{13.824 \times 2.4^2}{1} = 79.626 \ldots$ etc.

$\sqrt[3]{a}$. Apply Rule IV.

Ex. $\sqrt[3]{68} = 4.0816.$

$\sqrt[4]{a}$.   Take square root twice by use of lines *B* and *C*.

Ex. $\sqrt[4]{256}$: $\sqrt{256} = 16$, and $\sqrt{16} = 4$.

$\sqrt[6]{a}$.   Take cube root by Rule IV., then square root by use of lines *B* and *C*.

Ex. $\sqrt[6]{729}$: $\sqrt[6]{729} = 9$, and $\sqrt{9} = 3$ . . . etc.

FRACTIONS.

To Reduce Common to Decimal Fractions.

First—To reduce fractions having a common denominator to their equivalent decimal values.

$\dfrac{1 \times \text{numerator}}{\text{denominator}}$ = decimal value (see Rule I.)   Opposite the common denominator on *B* set 1 on *A*, then opposite each numerator on *B* read the value of fraction on *A*.

Ex. $\dfrac{7}{64} = .10937$, $\dfrac{11}{64} = .17185$, $\dfrac{29}{64} = .45312$,

$$\dfrac{47}{64} = .7344 \ldots \text{etc.}$$

Second—To reduce fractions having a common numerator to their equivalent decimal values.

$\dfrac{1 \times \text{numerator}}{\text{denominator}}$ = decimal value.   Reverse the slide and apply Rule I.

To Reduce Decimals to Common Fractions.

First—To reduce decimals to their equivalent fractions having a common denominator.

$\dfrac{\text{denominator} \times \text{decimal}}{1}$ = numerator (see Rule I.)   Opposite

1 on B set the common denominator on A, then opposite the decimal on B read the numerator of fraction on A.

Ex. $.23437 = \frac{15}{64}$, $.32812 = \frac{21}{64}$, $.7656 = \frac{49}{64}$

$$.8906 = \frac{57}{64} \quad . \quad . \quad . \text{ etc.}$$

Second—To reduce decimals to their equivalent fractions having a common numerator.

$\dfrac{1 \times \text{numerator}}{\text{decimal}} = \text{denominator}.$ Reverse the slide and apply Rule I.

PART III.

SPECIAL APPLICATIONS.

Conversion of Foreign and Domestic Weights and Measures.

Rule.—Opposite any quantity in one denomination on *B* set its value in terms of the other denomination on *A*, then opposite any other quantity of the first denomination on *B* will be found its value in terms of the other denomination on *A*, and *vice versa.*

Examples.

B. Granite, cubic feet . . . . Opp.   1   Opp.   8.5,   5.882 etc.
A.   "   lbs. avoir . . . . . Set   170   Find 1445, 1000   "

B. Cast-iron, cubic inches. . Opp.   1   Opp. 121,   768   "
A.   "   lbs. avoir . . . . Set .2604   Find 31.5,   200   "

B. French kilogrammes . . . Opp.   1   Opp. 25,   60   "
A. U. S. lbs. avoir . . . . Set 2.2047   Find 55.12, 132.28 "

B. Miles . . . . Opp.   1   Opp. 12,   2.84 "
A. Yards . . . . Set 1760   Find 21120, 5000 "

B. Acres . . . . . ·. . . . . Opp.   1   Opp.   5,   2.066 "
A. Square yards . . . . Set 4840   Find 24200, 10000 "

B. U. S. bushels . . . . Opp.   1   Opp. 2.5,   4.7 "
A. Cubic inches . . . . Set 2150   Find 5375, 10105 "

B. U. S. gallons . . . . Opp.   1   Opp.   4,   7.6 "
A. Cubic inches . . . . Set   231   Find 924, 1755.6 "

B U. S. gallons . . . . Opp.   1   Opp.   9,   18.007 "
A. Imperial gallons . . . . Set .833   Find 7.497, 15   "

34

B. French metres .... Opp. 1   Opp. 7,   30.48 etc.
A. U. S. feet............. Set 3.2807   Find 22.965, 100   "

B. French square metres.. Opp. 1   Opp. 2.5, 232.28 "
A. U. S. square feet ...... Set 10.763   Find 26.907, 2500   "

B. French cubic metres... Opp. 1   Opp. 5,   19.116 "
A. U. S. cubic yards...... Set 1.3078   Find 6.539,   25   "

The value of the instrument in questions of this nature, which are of constant occurrence in business, is beyond estimate. It can be applied to the exchange of money, all pro-rata questions, taxes, fellowship, pay-rolls, the mechanical powers and machinery, specific gravity, the conversion of thermometric scales, etc. —in fact, nearly all calculations are but questions in proportion, solved by the same general rules heretofore given. The letters on the left refer in all cases to the line or scale on which the quantities opposite are taken.

Exchange of Money.

B. British pounds sterling.. Opp. 1   Opp. 40,   24.658 etc.
A. U. S. dollars .... Set 4.8665   Find 194.66, 120   "

B. French francs .... Opp. 1   Opp. 250, 243.9   "
A. U. S. dollars .... Set .1927   Find 48.17,   47   "

B. German ducats .... Opp. 1   Opp. 50,   31.526 "
A. U. S. dollars .... Set 2.2838   Find 114.19, 72   "

Pro-Rata.

B	Opp. total of accounts	Opp. 1	Opp. each account
		or	then
A	Set amount divided	Set rate	Find am't going to it

Ex. $541.36 are to be divided pro-rata among various accounts amounting to $7436.00 ; required the amount going to account of $427.50, $763.80, etc.

B	Opp. 7436	Opp. 1	Opp. $427.50, $763.80, etc.
		or	then
A	Set 541.36	Set .0728	Find 31.12, 55.60 "

Taxes.

B	Opp. total valuation		Opp. 1		Opp. each valuation
		or		then	
A	Set amount raised		Set rate		Find amount of tax.

Ex. If $2362.50 are to be raised on a total valuation of $375000, required the tax on valuations of $2000, $4500, etc.

B	Opp. 375000		Opp. 1		Opp. $2000, $4500, etc.
		or		then	
A	Set 2362.50		Set .0063		Find 12.60, 28.35 "

Fellowship.

B	Opp. capital stock		Opp. 1		Opp. each share
		or		then	
A	Set total gain or loss		Set rate		Find profit or loss.

Ex. A capital stock of $75000 gave a gain of $4500 ; required the profit on shares of $600, $900, etc.

B	Opp. 75000		Opp. 1		Opp. $600.00, $900.00, etc.
		or		then	
A	Set 4500		Set .06		Find 36.00, 54.00 "

Pay-Rolls.
Daily Wages.

B	Opp. 1	Opp. number of days
A	Set rate per day	Find amount.

Ex. Required the amount, at $1.60 per day, for 9 days 7 hours, 16 days 3 hours, etc.

B	Opp. 1	Opp. 9.7, 16.3, etc.
A	Set 1.60	Find 15.52, 26.08 "

Weekly Wages.

B	Opp. 6	Opp. number of days
A	Set rate per week	Find amount.

Ex. Required the amount, at $10 per week, for 5 days, 6 days 3 hours, etc.

B	Opp. 76	Opp. 5, 6.3, etc.
A	Set 10	Find 8.33, 10.50 "

Monthly Wages.

B	Opp. working days in month	Opp. number of days
A	Set rate per month	Find amount.

Ex. Required the amount, at $65 per month of 26 days, for 10 days, 25 days 6 hours, etc.

B	Opp. 26	Opp. 10, 25.6, etc.
A	Set 65	Find 25.00, 64.00 "

The Mechanical Powers.

Any Order of Lever, or the Wheel and Axle.

B	Opp. distance of power from fulcrum	Opp. weight
A	Set " weight " "	Find power.

Any Combination of Wheels.

B	Opp. product of radii of wheels	Opp. weight
A	Set " " " pinions	Find power.

Any Combination of Pulleys.

B	Opp. number of pulleys	Opp. weight
A	Set 1	Find power.

Inclined Plane and Wedge.

B	Opp. length	Opp. resistance
A	Set height	Find power.

Screw.

B	Opp. circum. described by power	Opp. weight
A	Set distance c. to c. of threads	Find power.

In any of the above examples, the weight and power being known, the order of setting may be reversed and the unknowns found. About one-third more power than above given must be allowed to overcome friction.

Machinery.

Toothed Wheels.

B	Opp. 3.1416	Opp. number of teeth
A	Set pitch	Find diameter.

To find the diameters of two wheels to work at given actual or relative revolutions, the distance between centres being known.

B	Opp. sum of revolutions	Opp. number of revolutions
A	Set distance between centres	Find radius of other wheel.

Specific Gravity.

B	Opp. weight lost in water	Opp. 1000
A	Set total weight	Find specific gravity.

Velocity of Sound.

B	Opp. 1	Opp. time in seconds
A	Set 1090 (air) or 4708 (water)	Find distance in feet.

Uniform Motion.

B	Opp. 1	Opp. miles per hour
A	Set 1.4667	Find feet per second.

Conversion of Thermometric Scales.

			Above Freezing.	Bet. Freezing and o Fahr.	Below o Fahr.
B	Fahrenheit	Opp. 9	Deg.—32	32 — deg.	32 + deg.
A	Centigrade	Set 5	Deg.	Deg.	Deg.
B	Fahrenheit	Opp. 9	Deg.—32	32 — deg.	32 + deg.
A	Réaumur	Set 4	Deg.	Deg.	Deg.
B	Centigrade	Opp. 5	Deg.	Deg.	Deg.
A	Réaumur	Set 4	"	"	"

The preceding special applications make use of the lines *A* and *B*, and are all examples of the first of formulæ (1), or $\frac{ax}{b}$ = answer, in which $b : a :: x :$ answer. The statements have been made in terms of the proportion, or as follows:

B	Opp. *b* (first term)	Opp. *x* (third term)
A	Set *a* (second term)	Find answer (fourth term).

which are but applications of Rule I. to the formula $\frac{ax}{b}$.

It is more desirable in some respects to state a question by formulæ, as it expresses clearly and compactly the relation which the different parts bear to each other; and this will be done for the most part hereafter. As numerous examples have been given under the preceding rules, and as the learner is supposed to be familiar with them, it will not be necessary in the following examples to give special directions in each case for setting and reading.

Motion Uniformly Accelerated or Retarded by Gravity.

Height in feet

$$= \frac{16.083 \times (\text{time in seconds})^2}{1} = \frac{1 \times (\text{velocity in ft. per second})^2}{64.333}$$

Time in seconds $= \dfrac{1 \times (\text{velocity in ft. per second})}{32.166}$.

Ex. A stone is dropped from such a height as to require 3 seconds in falling.

Height $= \dfrac{16.083 \times 3^2}{1} = 144.75$ ft.

Velocity $= \dfrac{3 \times 32.166}{1} = 96.5$ ft. per second at end of third second.

The instrument is admirably adapted to the solution of all dynamic formulæ.

Pendulums.

Length in inches (New York) $= \dfrac{39.1 \times (\text{time in seconds})^2}{(\text{number of single oscillations})^2}$

Ex. Length required for 20 vibrations in 8 seconds $= \dfrac{39.1 \times 8^2}{20^2}$ $= 6.25$ inches.

Timber Measure.

Squared Timber.

$\dfrac{(\text{length in ft.}) \times (\text{area in sq. in.})}{144} =$ contents in cubic ft.

Ex. 1 piece $12'' \times 16'' \times 23'$ long $= \dfrac{23 \times 192}{144} = 30.67$ cubic ft.

Round Timber.

$\dfrac{(\text{length in ft.}) \times (\text{girth in inches})^2}{48^2} =$ contents in cubic ft.

Ex. 1 piece 40 ft. long and 50 in. in girth $= \dfrac{40 \times 50^2}{48^2} = 43.4$ cubic feet.

The above gives only about .8 of the true contents, and compensates for waste in squaring. The following gives the true contents very nearly :

$$\frac{(\text{length in ft.}) \times (\text{girth in inches})^2}{42.5^2} = \text{contents in cubic ft.}$$

Centrifugal Force of Railway Trains on Curves.

c = centrifugal force per lineal foot for each degree of curvature.
w = weight of train per lineal foot of track.
v = velocity of train in miles per hour.

then
$$c = \frac{wv^2}{85730}$$

Ex. A train weighing 2500 lbs. per lineal foot passes over a bridge on a 4° curve at a speed of 30 miles an hour ; required its effect upon the lateral system.

$$\frac{4 \times 2500 \times 30^2}{85730} = 105 \text{ lbs. per lineal foot of bridge.}$$

Mensuration.

Lines.

Diam. of circle

$$= \frac{1 \times \text{circum.}}{3.1416} = \frac{1 \times (\text{side of equal sq.})}{.8862} = \frac{1 \times (\text{side of insc. sq.})}{.7071}$$

Surfaces.

Area of circle $= \dfrac{.7854 \times \text{diam.}^2}{1} = \dfrac{.07958 \times \text{circum.}^2}{1}$

Area of ellipse $= \dfrac{(\text{major axis}) \times (\text{minor axis})}{1.2732}$

Area of parabola $= \dfrac{\text{chord} \times \text{height}}{1.5}$

Area of hexagon

$$= \frac{.866 \times (\text{diam. of insc'd circle})^2}{\text{I}} = \frac{.6495 \times (\text{diam. circ'd circ.})^2}{\text{I}}$$

Area of octagon $= \dfrac{.8284 \times (\text{diam. of insc'd circ.})^2}{\text{I}}$

$$= \frac{.7071 \times (\text{diam. of circ'd circ.})^2}{\text{I}}$$

Surface of sphere $= \dfrac{3.1416 \times \text{diam.}^2}{\text{I}}$

Convex surface of a cylinder $= \dfrac{\text{diam.} \times \text{height}}{.3183}$

Convex surface of a cone $= \dfrac{\text{diam.} \times \text{slant height}}{.6366}$

Solids.

Solidity of sphere $= \dfrac{\text{diam.} \times \text{diam.}^2}{1.9098}$

Solidity of cone $= \dfrac{\text{height} \times \text{diam.}^2}{3.8197}$

Solidity of cylinder $= \dfrac{\text{length} \times \text{diam.}^2}{1.2732}$

Many other lines, surfaces, and solids are worked with equal facility.

Cask-Gauging.

Casks are generally gauged by reducing them to four varieties, but Mr. Thomas Kentish proposes a substitute applicable to all, and which, tested by actual measurement of 50 casks aggregating 7400 gallons, gave a total error of only 8 gallons. It is as follows :

$C =$ Contents in gallons.
$L =$ Length in inches.
$H =$ Head diameter in inches.
$B =$ Bung diameter in inches.

$$C = \frac{L \times (H^2 + 2B^2)}{1089} = \text{imperial galls. of } 277.27 \text{ cu. in.}$$

$$C = \frac{L \times (H^2 + 2B^2)}{907.3} = \text{U. S. galls. of } 231 \text{ cu. in.}$$

Example:

$L = 28$ inches.
$H = 23$ "
$B = 28$ "

$$C = \frac{28(23^2 + 2 \times 28^2)}{907.3} = 16.32 + 2 \times 24.19 = 64.7 \text{ U. S. galls.}$$

Opposite 907.3 on B set 28 on A, then opposite 23 on C find 16.32 on A, and opposite 28 on C find 24.19 on A.

Ullage of Lying Casks.

$U =$ Ullage, or liquid in cask.
$W =$ Wet inches, or depth at bung.

$$U = \frac{C \times (10W - B)}{8B}$$

Example:

$W = 11$ inches.

Dimensions of cask as above,

$$U = \frac{64.7(110 - 28)}{224} = 23.68 \text{ U. S. gallons.}$$

Gauge-Points.

Various questions of a complex nature are worked with great facility by the use of previously calculated and tabulated constants called gauge-points, which are used for all similar questions.

These are found by combining the constants of any general formula, shifting them from the numerator to the denominator, or the reverse if necessary, by reciprocals, and thus reducing to one of the general forms applicable to the instrument.

The following applications will illustrate their time-saving properties.

Simple Interest.

$$\text{Int.} = \frac{\text{principal} \times \text{rate} \times \text{time in days}}{100 \times 360} = \frac{\text{principal} \times \text{time in days}}{g}$$

in which $g = \dfrac{36000}{\text{rate } \%}$

Rate %	g
4	9000
4½	8000
5	7200
6	6000
	etc.

By Rule I. Opposite g on B set principal on A, then opposite time in days on B will be found the interest on A.

Ex. $478.50 for 1 year, 2 months, and 14 days $=$ (434 days) at 4½ per cent. $= \dfrac{478.50 \times 434}{8000} = $25.96.

Opposite 8000 on B set 478.50 on A, then opposite 434 on B find 25.96 on A.

Ex. \$743.80 for 5 months and 16 days (166 days) at 6 per cent.
$$= \frac{743.80 \times 166}{6000} = \$20.58, \text{ etc.}$$

Bank Discount.

Bank discount is the same as simple interest, three days being allowed for grace.

Ex. The discount on \$1743.00, due in 60 days at 5 per cent.,
$$= \frac{1743 \times 63}{7200} = \$15.25.$$

The present worth $= 1743.00 - 15.25 = \$1727.75$.

True Discount.

$$\text{Present worth} = \frac{\text{amount due at maturity}}{1.00 + \frac{1 \times \text{time in days}}{g}}$$

Ex. Required the present worth of \$1743.00, due in 60 days at 5 per cent., three days being allowed for grace.

$$\text{Present worth} = \frac{1743.00}{1.00 + \frac{1 \times 63}{7200}} = \frac{1743.00}{1.00875} = \$1727.90.$$

Discount $= 1743.00 - 1727.90 = \$15.10$.

Questions of this nature, as is evident from the formula, require two applications of Rule I. In the above example—

First—Opposite 7200 on B set 1 on A, then opposite 63 on B find .00875 on A. (See Rule II.)

Second—Opposite 1.00875 on B set 1 on A, then opposite 1743 on B find 1727.90 on A.

Compound Interest.

$$\text{Amount} = \frac{\text{principal} \times \left(1.00 + \frac{\text{rate}}{100}\right)^{n}}{1} = \frac{\text{principal} \times g}{1} \text{ in which}$$

g = principal and interest of \$1.00 for the given rate and time.

Values of g at 4, 4½, 5, and 6 per cent., for from 1 to 10 years, are given in the table.*

	Values of g.			
Years.	Rate per cent.			
	4	4½	5	6
1	1.04	1.045	1.05	1.06
2	1.0816	1.0921	1.1025	1.1236
3	1.1248	1.1411	1.1576	1.191
4	1.1698	1.1925	1.2155	1.2625
5	1.2166	1.2462	1.2763	1.3382
6	1.2663	1.3023	1.3401	1.4185
7	1.3159	1.3609	1.4071	1.5033
8	1.3685	1.4221	1.4771	1.594
9	1.4233	1.4861	1.5513	1.6895
10	1.4802	1.5530	1.6289	1.791

* The table may be extended to any number of years, at any desired rate per cent., mechanically as follows :

Remove the slide from the envelope. With a pair of dividers place one leg at 100, the commencement of scale A, and the other leg at 100+rate per cent.; for 3 per cent., for instance, the dividers would extend from 100 to 103. Lay off this space continuously on scale A as many times as the required number of years ; the reading of the scale at the end of any number of spaces is the value of g for that number of years. Care must be taken that the reading at the last point on any line is made the starting-point on the following line.

Ex. What will be the amount of $250.00 placed at compound interest for 10 years at 6 per cent.?

From table, $g = 1.791$ ∴ Amount $= \dfrac{250.00 \times 1.791}{.1} = \$447.75.$

Ex. $300.00 for 6 years at 4½ per cent.?

$g = 1.3023$ ∴ Amount $= \dfrac{300.00 \times 1.3023}{1} = \$390.69.$

Weights of Materials.

Prisms.

Weight $= \dfrac{\text{length} \times \text{area}}{g}$ in which g is the reciprocal of the weight of a unit of volume.

Length in ft., area in ft., $g =$ reciprocal of weight of 1 cu. ft.
    "   "   " in., "   "   "   " 12 cu. in.
  "   in.,   " in., "   "   "   1 "

Cylinders:

Weight $= \dfrac{\text{length} \times \text{diam.}^2}{g}$

Length in ft., diameter in ft., $g =$ reciprocal of weight of cylinder
    1 ft. long and 1 ft. diameter.
Length in ft., diameter in in., $g =$ reciprocal of weight of cylinder
    1 ft. long and 1 in. diameter.
Length in in., diameter in in., $g =$ reciprocal of weight of cylinder
    1 in. long and 1 in. diameter.

Spheres.

Weight $= \dfrac{\text{diam.} \times \text{diam.}^2}{g}$

Diameter in ft., $g =$ reciprocal of weight of sphere 1 ft. in diam.
    "   in., "   "   "   "   " 1 in.   "

Cones.

The values of g are three times the corresponding ones for cylinders.

The following table gives the values of g for prisms, cylinders, and spheres of materials in common use. F and I denote feet and inches.

MATERIAL.	PRISM.						CYLINDER.						SPHERE.		Weight of 1 cu. ft.
	Length.	Area.	Length.	Area.	Length.	Area.	Length.	Diam.	Length.	Diam.	Length.	Diam.	Diam.	Diam.	
	F	F	F	I	I	I	F	F	F	I	I	I	F	I	w
Wrought-iron	.00208	.300	3.60				.00265	.382	4.58				.00398	6.87	480
Cast-iron ...	.00222	.320	3.84				.00283	.407	4.89				.00424	7.33	450
Steel	.00204	.294	3.53				.00260	.374	4.49				.00390	6.73	490
Brass.......	.00191	.275	3.30				.00243	.350	4.20				.00364	6.30	524
Copper	.00183	.263	3.16				.00232	.335	4.02				.00348	6.02	548
Lead	.00141	.203	2.44				.00179	.258	3.09				.00269	4.94	711
Sandstone ...	.00662	.954	11.45				.00843	1.214	14.57				.01265	21.85	151
Limestone ...	.00595	.857	10.28				.00758	1.091	13.10				.01137	19.64	168
Granite......	.00588	.847	10.16				.00749	1.078	12.94				.01123	19.41	170
White pine ..	.0333	4.80	57.60				.0424	6.111	73.34				.06366	110.0	30
Yellow pine. .	.0250	3.60	43.20				.0318	4.584	55.00				.04774	82.5	40
White oak....	.0189	2.72	32.64				.0240	3.459	41.51				.03604	62.3	53
Any material.	$g=\frac{1}{w}$	$\frac{144}{w}$	$\frac{1728}{w}$				$\frac{1.2732}{w}$	$\frac{183.34}{w}$	$\frac{2200}{w}$				$\frac{1.9098}{w}$	$\frac{3300}{w}$	

Examples in Use of Table.

Block of sandstone 4 ft. long, 3 ft. wide, and 1.5 ft. high.

$$\text{Weight} = \frac{4 \times 4.5}{.00662} = 2719 \text{ lbs.}$$

Bar of wrought-iron 26.5 ft. long and 6.4 sq. in. area.

$$\text{Weight} = \frac{26.5 \times 6.4}{.300} = 565.3 \text{ lbs.}$$

Bar of steel 27 in. long and 5.0 sq. in. area.

$$\text{Weight} = \frac{27 \times 5.0}{3.53} = 38.2 \text{ lbs.}$$

Cylinder of brass 16 ft. long and 1.3 ft. diam.

$$\text{Weight} = \frac{16 \times 1.3^2}{.00243} = 11127 \text{ lbs.}$$

Hollow cylinder of copper 12 ft. long and $\begin{vmatrix} 12'' \text{ outside} \\ 11'' \text{ inside} \end{vmatrix}$ diam.

$$\text{Weight} = \frac{12 \times (12^2 - 11^2)}{.335} = 824 \text{ lbs.}$$

Cylinder of lead 23 in. long and 4¼ in. diam.

$$\text{Weight} = \frac{23 \times 4.25^2}{3.09} = 134.4 \text{ lbs.}$$

Globe of granite 2.5 ft. diam.

$$\text{Weight} = \frac{2.5 \times 2.5^2}{.01123} = 1391 \text{ lbs.}$$

Ball of cast-iron 7.5 in. diam.

$$\text{Weight} = \frac{7.5 \times 7.5^2}{7.33} = 57.5 \text{ lbs.}$$

Weight of any Number of Pieces of Wrought-Iron or Steel.

The weight of one piece of any material, of any required section, may be readily found by the use of gauge-points of the character heretofore described; but for some materials, like wrought-iron and steel, which are used to a large extent, it will be found of great advantage to make a table giving the gauge-point for each number of pieces from 1 to 100. It may be made with reference either to a given area or a given weight per lineal foot.

By this means the weight of any number of pieces of any length, and of any area or weight per foot, may be found by one operation. For wrought-iron a table having reference to a given weight per lineal foot will be found the most useful, for shape-iron is usually ordered in this way, and the weight per foot of any area of bar-iron can be found mentally by adding a cipher and dividing by three.

Given the number of pieces, length, and weight per foot,

$$\text{total weight} = \frac{\text{length in ft.} \times \text{weight per ft.}}{g}$$ in which g is the

reciprocal of the number of pieces.

Reciprocals of numbers from 1 to 100 are as follows :

No. Pcs.	0	1	2	3	4	5	6	7	8	9
0	....	1.0000	.50000	.33333	.25000	.20000	.16667	.14286	.12500	.11111
10	.10000	.09091	.08333	.07692	.07143	.06667	.06250	.05882	.05555	.05263
20	.05000	.04762	.04545	.04348	.04167	.04000	.03846	.03704	.03571	.03448
30	.03333	.03226	.03125	.03030	.02941	.02857	.02778	.02703	.02632	.02564
40	.02500	.02439	.02381	.02326	.02273	.02222	.02174	.02128	.02083	.02041
50	.02000	.01961	.01923	.01887	.01852	.01818	.01786	.01754	.01724	.01695
60	.01667	.01639	.01613	.01587	.01562	.01538	.01515	.01492	.01471	.01449
70	.01429	.01408	.01389	.01370	.01351	.01333	.01316	.01299	.01282	.01266
80	.01250	.01235	.01220	.01205	.01190	.01176	.01163	.01149	.01136	.01124
90	.01111	.01099	.01087	.01075	.01064	.01053	.01042	.01031	.01020	.01010

Examples.

The weight of 16 pieces of wrought-iron 17.4 ft. long and 19.7 lbs. per ft. $= \dfrac{17.4 \times 19.7}{.0625} = 5484$ lbs.

The weight of 84 pieces of wrought-iron 14.2 ft. long and 9.6 lbs. per ft. $= \dfrac{14.2 \times 9.6}{.0119} = 11455$ lbs. . . . etc.

Given the number of pieces, length, and area of section, total weight $= \dfrac{\text{length in ft.} \times \text{area in sq. in.}}{g}$ in which g is the reciprocal of weight of bar 1 ft. long and 1 in. sq., divided by the number of pieces.

For steel, $g = \dfrac{.2941176}{\text{No. pcs.}}$

Values of g for steel, up to 60 pieces, are as follows:

No. Pcs.	0	1	2	3	4	5	6	7	8	9
0	....	.29412	.14705	.09804	.07353	.05882	.04902	.04202	.03676	.03268
10	.02941	.02674	.02451	.02262	.02101	.01961	.01838	.01730	.01634	.01548
20	.01471	.01401	.01337	.01279	.01225	.01176	.01131	.01089	.01050	.01014
30	.009804	.009486	.009189	.008908	.008649	.008401	.008168	.007948	.007736	.007539
40	.007353	.007173	.007001	.006839	.006683	.006535	.006393	.006257	.006126	.006001
50	.005882	.005766	.005655	.005548	.005445	.005347	.005252	.005159	.005069	.004984

Examples.

The weight of 28 pieces of steel 24.5 ft. long and 8.4 sq. in. area $= \dfrac{24.5 \times 8.4}{.0105} = 22400$ lbs.

The weight of 9 pieces of steel 35.4 ft. long and 6.25 sq. in. area $= \dfrac{35.4 \times 6.25}{.03268} = 6770$ lbs

Weight of Sheet Metals.

Total weight $= \dfrac{\text{length in in.} \times \text{width in in.}}{g}$

For 1 piece, $g = \dfrac{144}{\text{weight per sq. ft.}}$

For any number of pieces, $g = \dfrac{\left(\dfrac{144}{\text{weight per sq. ft.}}\right)}{\text{No. pieces.}}$,

Values of g up to 10 pieces :

No. Pieces.	Number of Birmingham or Stubs' Wire-gauge.*						
	16	18	20	22	24	26	28
1	55.385	73.469	102.86	128.57	163.64	200.0	257.14
2	27 69	36.73	51.43	64.28	81.82	100.0	128.6
3	18.46	24.49	34.29	42.86	54.55	66.67	85.71
4	13.85	18.37	25.71	32.14	40.91	50.00	64.28
5	11.08	14.69	20.57	25.71	32.73	40.00	51.43
6	9.231	12.24	17.14	21.43	27.27	33.33	42.86
7	7.912	10.50	14.69	18.37	23.37	28.57	36.73
8	6.923	9.183	12.86	16.07	20.45	25.00	32.14
9	6.154	8.163	11.43	14.28	18.18	22.22	28.57
10	5.538	7.347	10.29	12.86	16.36	20.00	25.71

The above table, as well as all preceding ones, may be rapidly extended to any number of pieces, and for any gauge, by aid of the instrument. As the denominator is the variable, reverse the slide and apply Rule I. One setting only is required for each column.

Examples.

The weight of 9 sheets of wrought-iron $31'' \times 66''$, No. 20 B. G
$= \dfrac{31 \times 66}{11.43} = 179$ lbs.

* A cubic foot of wrought-iron is assumed to weigh 480 lbs.

The weight of 7 sheets of wrought-iron $27'' \times 42''$, No. 16 B. G.
$= \dfrac{27 \times 42}{7.912} = 1433$ lbs.

Board Measure.

Feet B. M. $= \dfrac{\text{length in. ft.} \times \text{sec. in sq. in.}}{g}$ in which $g = \dfrac{12}{\text{No. pcs.}}$

Values of g up to 60 pieces :

No. Pcs.	0	1	2	3	4	5	6	7	8	9
0	....	12.000	6.0000	4.0000	3.0000	2.4000	2.0000	1.7143	1.5000	1.3333
10	1.2000	1.0909	1.0000	0.9220	0.8572	0.8000	0.7500	0.7058	0.6667	0.6316
20	0.6000	0.5714	0.5454	0.5218	0.5000	0.4800	0.4615	0.4444	0.4285	0.4138
30	0.4000	0.3871	0.3750	0.3636	0.3529	0.3428	0.3333	0.3244	0.3157	0.3077
40	0.3000	0.2927	0.2857	0.2790	0.2726	0.2666	0.2609	0.2554	0.2500	0.2449
50	0.2400	0.2353	0.2308	0.2264	0.2222	0.2182	0.2143	0.2105	0.2069	0.2034

Examples.

24 pieces $7'' \times 14'' \times 26'$ long $= \dfrac{98 \times 26}{.5} = 5096'$ B. M.

Opposite 5 on B set 98 on A, then opposite 26 on B find 5096 on A.

47 pieces $7'' \times 9'' \times 16.5'$ long $= \dfrac{63 \times 16.5}{.2554} = 4070'$ B. M., etc.

Measures of Capacity.

Dimensions are given in feet, inches, or both, and the cubic contents are required in inches, feet, yards, perches, gallons, bushels, etc.

The gauge-point of any unit of measure is the value of that unit expressed in terms of the unit of the given dimensions.

Prisms.

$$\text{Contents} = \frac{\text{length} \times \text{area}}{g}$$

Length in ft., area in ft., $g =$ number of cubic ft. contained in 1 unit of measure.

Length in ft., area in in., $g =$ number of 12 cubic in. contained in 1 unit of measure.

Length in in., area in in., $g =$ number of cubic in. contained in 1 unit of measure.

Cylinders.

$$\text{Contents} = \frac{\text{length} \times \text{diam.}^2}{g}$$

Length in ft., diam. in ft., $g =$ number of cylinders 1 ft. long and 1 ft. diam. contained in 1 unit of measure.

Length in ft., diam. in in., $g =$ number of cylinders 1 ft. long and 1 in. diam. contained in 1 unit of measure.

Length in in., diam. in in., $g =$ number of cylinders 1 in. long contained in 1 unit of measure.

Spheres.

$$\text{Contents} = \frac{\text{diam.} \times \text{diam.}^2}{g}$$

Diam. in ft., $g =$ number of spheres 1 ft. in diam. contained in 1 unit of measure.

Diam. in in., $g =$ number of spheres 1 in. in diam. contained in 1 unit of measure.

Cones.

The values of g are three times the corresponding ones for cylinders.

The following table gives the values of g for prisms, cylinders, and spheres adapted to measures in common use. F and I denote feet and inches :

Measures.	Prism.						Cylinder.						Sphere.	
	Length.	Area.	Length.	Area.	Length.	Area.	Length.	Diam.	Length.	Diam.	Length.	Diam.	Diam.	Diam.
	F	F	F	I	I	I	F	F	F	I	I	I	F	I
Cubic inch ...	.000579	.0833		1			.000737	.1061	1.2732				.001106	1.9098
Cubic foot.....	1		144		1728		1.2732	183.3	2200				1.9098	3300
Cubic yard	27.		3888		46656		34.38	4950	59402				51.56	89100
Perch of stone.	24.75		3564		42768		31.51	4538	54452				47.27	81680
U. S. gallon...	.1337		19.25		231		.1702	24.51	294.11				.2553	441.2
Imperial gallon	.1605		23.11		277.3		.2043	29.42	353.06				.3065	529.6
U. S. bushel...	1.244		179.2		2150.4		1.584	228.1	2737.9				2.376	4107

Examples in Use of Table.

A bin 8.0 ft. long, 6.0 ft. wide, and 4.5 ft. high contains $\frac{8 \times 27}{1.244} = 173.6$ U. S. bushels.

A trough 12 ft. long, 18 in. wide, and 12 in. high contains $\frac{12 \times 216}{23.11} = 112.16$ imperial gallons.

A stone 50 in. long, 26 in. wide, and 14 in. high contains $\frac{50 \times 364}{1728} = 10.53$ cubic feet.

A cistern 12 ft. deep and 6.5 ft. in diam. contains $\frac{12 \times 6.5^2}{.1702} = 2979$ U. S. gallons.

A column 25 ft. high and 18.5 in. in diam. contains $\frac{25 \times 18.5^2}{4538} = 1.885$ perches.

A cylinder 27 in. long and 9.5 in. in diam. contains $\dfrac{27 \times 9.5^2}{1.2732}$

$= 1914$ cubic inches.

A sphere 16 ft. in diam. contains $\dfrac{16 \times 16^2}{51.56} = 79.44$ cubic yards.

A sphere 18 in. in diam. contains $\dfrac{18 \times 18^2}{441.2} = 13.22$ U. S. gallons.

Beams.

$M =$ Maximum bending moment in ft. lbs.
$L =$ Length in feet.
$w =$ Uniform load per lineal ft. in lbs.
$W =$ Concentrated load at centre.
W_1, W_2, W_3, etc. $=$ Loads at any points.
x_1, x_2, x_3, etc. $=$ Distances from nearest support.
$I =$ Moment of inertia of section.
$S =$ Strain per sq. in. in outer fibre.
$t =$ Distance of centre of gravity from outer fibre.
$b =$ Breadth of section in inches.
$d =$ Depth of section in inches.

Maximum Bending Moments.

Fixed at one end and loaded at the other.. $M = WL$ (1)

Fixed at one end and uniformly loaded. ... $M = wL^2 \div 2$ (2)

Supported at both ends and loaded at centre. $M = WL \div 4$ (3)

Supported at both ends and uniformly
 loaded .... $M = wL^2 \div 8$ (4)

Fixed at one end, supported at the other,
 and loaded at centre................. $M = 3 WL \div 16$ (5)

Fixed at one end, supported at the other,
 and uniformly loaded .... $M = wL^2 \div 8$ (6)

Fixed at both ends and loaded at centre... $M = WL \div 8$ (7)

Fixed at both ends and uniformly loaded.. $M = wL^2 \div 12$ (8)

Supported at both ends and loaded in
 any manner, $M = (W_1 x_1 + W_2 x_2 + W_3 x_3,$ etc.$) \div 2$ (centre) (9)

Strength.

Any section	$M = SI \div 12t$	(10)
Symmetrical section	$M = SI \div 6d$	(11)
Rectangular section	$M = Sbd^2 \div 72$	(12)

WOODEN BEAMS.

The moduli of rupture, or the ultimate values of S in the above formulæ, according to the experiments of Capt. Rodman, United States Engineer, are in round numbers as follows :

For yellow pine, white oak, ash, birch, and beech	10000
" white pine, poplar, and hemlock	7000
" spruce, cypress, and chestnut	6000

A factor of safety of about 6 for buildings and 8 for bridges is customary.

From (12) $b = \dfrac{72M}{Sd^2}$

$$\text{If } S = 900, \quad b = \frac{.08M}{d^2} \qquad (13)$$

$$\text{If } S = 1000, \quad b = \frac{.072M}{d^2} \qquad (14)$$

$$\text{If } S = 1200, \quad b = \frac{.06M}{d^2} \qquad (15)$$

Ex. A beam projecting 5 ft. from a wall carries a weight of 4000 lbs. at the free end. Required its size, allowing a fibre strain of 900 lbs. per sq. in.

From (1), $M = WL = 4000 \times 5 = 20000$ ft. lbs.

From (13), $b = \dfrac{.08M}{d^2}$ If $d = 12$ in., $b = \dfrac{.08 \times 20000}{12^2}$

$$= 11.1 \text{ in.} \quad \text{(Rule I.)}$$

From (13), $d = \sqrt{\dfrac{.08M}{b}}$ If $b = 9$ in., $d = \sqrt{\dfrac{.08 \times 20000}{9}}$

$$= 13.3 \text{ in.} \quad \text{(Rule III.)}$$

Ex. A beam 16.3 ft. between supports sustains a uniform load of 620 lbs. per lin. ft. Required its size, allowing a fibre strain of 1000 lbs. per sq. in.

$$\text{From (4),} \quad M = \frac{620 \times 16.3^2}{8} = 20591 \text{ ft. lbs.}$$

$$\text{From (14), If } d = 13 \text{ in., } b = \frac{.072 \times 20591}{13^2} = 8.77 \text{ in.}$$

(Rule I.)

Ex. A lintel over a 11 ft. opening has each end firmly embedded in a brick wall, and sustains a weight of 1000 lbs. per lin. ft. What depth is required for a width of 12 in. and a fibre strain of 1200 lbs. per sq. in.?

$$\text{From (8),} \quad M = \frac{1000 \times 11}{12} = 10083 \text{ ft. lbs.} \quad \text{(Rule I.)}$$

$$\text{From (15),} \quad d = \sqrt{\frac{.06 \times 10083}{12}} = 7.1 \text{ in.} \quad \text{(Rule III.)}$$

IRON BEAMS.

The working values of S in formulæ 10, 11, and 12 may be taken as follows :

For bridges, $S =$ from 8000 to 10000 lbs. per sq. in.
 " buildings, $S =$ " 12000 to 14000 "   "   "

The values of I for rolled beams, channels, etc., will be found in books published by the manufacturers of such shapes, and will not be given here, except such as are used in the following examples illustrating the application of the instrument to this class of work, and which correspond to the rolls of the Union Iron Mills, of Pittsburgh, Pa.

For 15″ "I" beams, 50 lbs. per ft., $I = 530$. For each additional pound up to 65, $I = 5.60$.
For 12″ "I" beams, 42 lbs. per ft., $I = 275$. For each additional pound up to 60, $I = 3.61$.

Ex. A 15″ "I" beam, resting upon supports 14.5 ft. apart, sustains a load of 17500 lbs. at centre. What weight of beam is required, if $S = 10000$ lbs. per sq. in. ?

From (3), $M = \dfrac{17500 \times 14.5}{4} = 63440$ ft. lbs. $\left.\vphantom{\dfrac{6dM}{S}}\right\}$ (Rule I.)

From (11), $I = \dfrac{6dM}{S} = \dfrac{90 \times 63440}{10000} = 571.$

Required weight per ft. $= 50 + \dfrac{(571 - 530)}{5.6} = 57.3$ lbs.

Ex. 12″ "I" beams, 5 ft. centre to centre and 20 ft. between supports, are used to support a load of 200 lbs. per sq. ft., or 1000 lbs. per lineal ft., of beam, including their own weight. What weight of beam is required, if $S = 12000$ lbs. per sq. in.?

From (4), $M = \dfrac{1000 \times 20^2}{8} = 50000$ ft. lbs. $\left.\vphantom{\dfrac{6dM}{S}}\right\}$ (Rule I.)

From (11), $I = \dfrac{6dM}{S} = \dfrac{72 \times 50000}{12000} = 300.$

Required weight per ft. $= 42 + \dfrac{300 - 275}{3.61} = 48.9$ lbs.

Ex. If beams weighing 42 lbs. per lin. ft. are used, what is the required distance, centre to centre, to support the above load?

From (11), $M = \dfrac{SI}{6d} = \dfrac{12000 \times 275}{72} = 45830$ ft. lbs. $\left.\vphantom{\dfrac{8M}{L^2}}\right\}$ (Rule III.)

From (4), $w = \dfrac{8M}{L^2} = \dfrac{8 \times 45830}{20^2}$

$\qquad = 916.6$ lbs. $\div 200 = 4.58$ ft. c. to c.

Example of Calculation for Plate-Girder Bridge.

$L =$ Span $= 62.0$ ft, c. to c. of bearings.
$d =$ Depth $= 5.0$ ft. of web $=$ effective depth nearly.*

* The effective depth is about ½ inch more at centre and ⅛ inch less at end than the depth of web, but the latter depth is sufficiently near for practical purposes.

$w =$ Dead-load $= 770$ lbs. per lineal ft. of bridge.

$w' =$ Rolling load $= \begin{cases} \text{80-ton consolidation engine} + \text{10\% impact.} \\ \text{equivalent to 3740 lbs. per lin. ft. of bridge.} \end{cases}$

$S = \begin{cases} \text{Strain per sq. in. in top flange} = 8000 \text{ lbs. (gross).} \\ \phantom{\text{Strain}} \text{``} \qquad \text{``} \qquad \text{`` bottom ``} = 10000 \text{ `` (net).} \end{cases}$

Rivets $\begin{cases} \text{Bearing, 15000 lbs. per sq. in. on diameter of hole.} \\ \text{Shearing, 7500 ``} \qquad \text{``} \qquad \text{``} \quad \text{area} \qquad \text{``} \end{cases}$

$(w + w') = 385 + 1870 = 2255$ lbs. per lin. ft. of girder.

Flange strain at centre $= \dfrac{(w + w') L^2}{8d} = \dfrac{2255 \times 62^2}{40} = 216700$ lbs.

(Rule I.)

($\frac{1}{8}$ area of web is available as flange area in each flange.)

Top flange, 216700 ÷ 8000 = 27.09 sq. in.	2 ang. $3\frac{1}{2}'' \times 5'' = 6.84$ sq. in.
`` Length = $62 \sqrt{\dfrac{16.5}{27.09}} = 48'.4$	1 plt. $12'' \times \frac{3}{8}'' = 4.50$ ``
``    `` $62 \sqrt{\dfrac{12.0}{27.09}} = 41'.3$	1 plt. $12'' \times \frac{1}{2}'' = 6.00$ ``
``    `` $62 \sqrt{\dfrac{6.0}{27.09}} = 29'.2$	1 plt. $12'' \times \frac{1}{2}'' = 6.00$ ``
	$\overline{}$ $\frac{1}{8}$ web.
	$= 23.34 + 3.75$
	gross.
	$= 27.09$ sq. in.
Bot. flange, 216700 ÷ 10000 = 21.67 sq. in.	2 ang. $3\frac{1}{2}'' \times 5'' = 6.84$ sq. in.
`` Length = $62 \sqrt{\dfrac{14.25}{24.84}} = 47'.0$	1 plt. $12'' \times \frac{5}{16}'' = 3.75$ ``
``    `` $62 \sqrt{\dfrac{10.5}{24.84}} = 40'.3$	1 plt. $12'' \times \frac{7}{16}'' = 5.25$ ``
``    `` $62 \sqrt{\dfrac{5.25}{24.84}} = 28'.5$	1 plt. $12'' \times \frac{7}{16}'' = 5.25$ ``
	$\overline{}$ $\frac{1}{8}$ web. rivet.
	$= 21.09 + 3.75 - 3.20$
	net.
	$= 21.64$ sq. in.
Web..........	1 plt. $60'' \times \frac{3}{8}'' = 22.50$ ``

The above formulæ for the length of plates explain themselves. They are derived from the equation of the parabola, and give the theoretic lengths, or lengths between points of intersection with a parabolic curve of moments. The first plates, top and bottom, are usually made full length for appearance' sake, and, in case of the top plate in deck-girders, to furnish a better seat for the cross-ties. For the remaining plates, the above lengths may be taken as the distance between extreme rivets, except when acting as covers to joints in web or preceding plates, in which case proper allowance must be made.

Length of plates by Rule III.:

Top flange $\begin{cases} \text{Opposite 62 on } C \text{ set 27.09 on } A, \text{ then opposite 16.5} \\ \text{on } A \text{ find 48.4 on } C; \text{ opposite 12.0 on } A \text{ find 41.3} \\ \text{on } C, \text{ and opposite 6.0 on } A \text{ find 29.2 on } C. \end{cases}$

Bottom flange $\begin{cases} \text{Opposite 62 on } C \text{ set 24.84 on } A, \text{ then opposite} \\ \text{14.25 on } A \text{ find 47.0 on } C; \text{ opposite 10.5 on} \\ A \text{ find 40.3 on } C, \text{ and opposite 5.25 on } A \text{ find} \\ \text{28.5 on } C. \end{cases}$

Shear per foot-run $\begin{cases} \text{End} = \dfrac{(w+w')L}{2d} = \dfrac{2255 \times 62}{10} = 13980 \text{ lbs.} \times \dfrac{23.34}{27.09} \\ \qquad\qquad = 12040 \text{ lbs. on rivets.} \\[2ex] \text{Centre} = \dfrac{w'L}{8d} = \dfrac{1870 \times 62}{40} = 2898 \text{ lbs.} \times \dfrac{23.34}{27.09} \\ \qquad\qquad = 2497 \text{ lbs. on rivets.} \end{cases}$

Rivets.

If F = shear per foot-run on rivets, and V = the bearing or shearing value of a rivet, the pitch or distance, centre to centre, of rivets = $12 V \div F$.

The bearing value of a $\frac{3}{4}''$ rivet, $\frac{13}{16}''$ hole, for $\frac{3}{8}''$ web = $.937 \times .375 \times 15000 = 5270$ lbs.

Pitch at ends = $12 \times 5270 \div 12040 = 5.2$ in., which may be increased towards the centre inversely as the shear up to the allowed maximum, usually 6 in.*

Vertical Stiffeners.

Rankine assumes that the shear per foot-run of girder (F_1) should not exceed $\dfrac{96000t}{1+\frac{1}{3000}r^2}$, t being the thickness of web in inches, and r the ratio of length to diameter of a section of web taken at an angle of 45° between horizontal or vertical angles; but if the least distance between such angles is substituted the result will be undoubtedly safe and more in keeping with ordinary practice, making this substitution

$$F_1 = 96000\,t \div \left(1 + \frac{r^2}{3000}\right),\ \text{or}\ r = \sqrt{\frac{288,000,000\,t}{F_1} - 3000}$$

for the present example,

$$r = \sqrt{\frac{288,000,000 \times .375}{13980} - 3000} = 68.7 \text{ at ends.}$$

∴ Horizontal distance in clear between stiffeners at ends should not exceed $68.7 \times \frac{3}{8} = 25.8$ in., which is increased towards the centre inversely as the shear.* If the required distance exceeds the clear distance between horizontal angles, no stiffeners are required.

The above radical is readily worked out by aid of the instrument as follows :

First find the value under the $\sqrt{}$ by Rule I., and then the root by use of lines B and C.

Compression Formulæ.

The formulæ used in the following estimate, and which are in more general use probably than any other, may be expressed as follows :

---

* For inverse or reciprocal proportion proceed as follows :

If $b : a$ inversely $:: c : d$.

Opposite b on B set a on A, then opposite c on A find d on B.

$$S = \frac{f}{1 + c\frac{l^2}{r^2}} = \frac{8000}{1 + .004\frac{l^2}{r^2}} \quad \text{flat at both ends}$$

$$= \frac{8000}{1 + .006\frac{l^2}{r^2}} \quad \text{flat at one end}$$

$$= \frac{8000}{1 + .008\frac{l^2}{r^2}} \quad \text{pin at both ends}$$

in which $S=$allowed strain per square inch, $c=$constant depending upon the condition of ends, $l=$length in feet, and $r=$ radius of gyration in inches.

As any single value of S requires two settings of the instrument, and any number of values for the three conditions requires but four, it is of advantage, if the formulæ are extensively applied, to first make a table giving the values of S for all values of $\frac{l^2}{r^2}$ from 1 to 500 or upwards. First, by Rule I., with slide direct, find all values of $c\frac{l^2}{r^2}$, which requires one setting for each value of c, then reverse the slide and by one additional setting find all values of S.

Connecting-Plates, Bearing-Plates, etc.

Some engineers require that the plates which connect the ends of the two channels or segments of which compression members are usually composed shall be provided with rivets sufficient to transmit one-half the strain on the member. This may be stated as follows :

Length of tie-plate
$$= \frac{\text{strain in member} \times \text{pitch of rivets}}{4 \times \text{value of rivet}} = \frac{\text{strain in member}}{g}$$

in which g is constant so long as the value and pitch of rivets remain constant, and the required lengths of all connecting-plates are found by one setting of the instrument.

The thickness and length of bearing-plates and other details in construction are found with the same facility.

Example of Truss-Bridge Calculation (Pratt Truss).

Fig. 6.

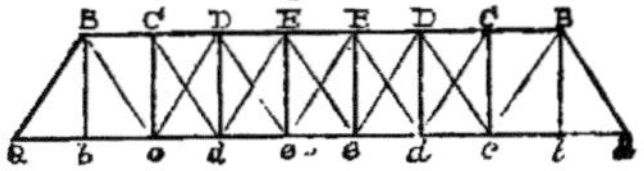

9 panels of 17.0 ft. each.

Span, 153' o.c. to c. of pins.   Depth, 25' o.c. to c. of chords.   Width, 16' o.c. to c. of trusses.

Dead-load, 1200 lbs. per lineal ft. bridge.   Rolling load, 2750 lbs. per lineal ft. bridge.

W = Dead-load per panel per truss = $600 \times 17 = 10200$ lbs. (⅓ at top and ⅔ at bottom).

W' = Rolling " " = $1375 \times 17 = 23375$ " (at bottom).

Max. " " = $1820 \times 17 = 30940$ " (for 74-ton consolidation engine).

$(W + W')$ Horizontal comp. = $33575 \times 17 \div 25 = 22831$ " (uniform).

Braces, $l = \sqrt{25.0^2 + 17.0^2} = 30.232$. $R = 30.232 \div 25 = 1.209$.

Wind $\begin{cases}\text{Top}\\ \text{Bot.}\end{cases}$ laterals per panel (uniform) = $150 \times 17 = 2550$ lbs. $R = 1.382.$
" " = $150 \times 17 = 2550$ "
" " (rolling) = $300 \times 17 = 5100$ " $\Big\}$ $R = 1.458.$

(But one setting of the instrument is required for each panel load or value of R.)

Member.	Uniform Load. Panel Load.	Mult.	Result.	Rolling Load. Panel Load.	Mult.	Result.	Imp. %	Max. Vert. Comp.	R	Max. Strain.	Allowed Strain per sq. inch.	Area Req'd.	Section Used.	Actual Area.
Web.														
aH	10200	4.0	40800	23375		93500	—	134300	1.209	162370	6440	25.21	2 15" chans.	25.20
Br	"	3.0	30600	"		72720	—	103320	"	124910	10000	12.49	2 bars 5"×1¼"	12.50
Cd	"	2.0	20400	"		54540	—	74940	"	90600	"	9.06	2 " 4"×1⅛"	9.00
De	"	1.0	10200	"		28960	10	53060	"	64150	"	6.41	2 " 4"×⅞"	6.50

Fe	10200	0.0	0	23375	$\frac{10}{9}$	25970	25	32460	1.209	39240	10000	3.92	2 bars $1\frac{3}{8}''$ sq.	3.78
Ed	“	−1.0	−10200	“	$\frac{6}{9}$	15580	25	9280	“	11220	“	1.12	2 “ $\frac{7}{8}''$ “	1.54
Dc	“	−2.0	−20400	“	$\frac{3}{9}$	7790	25	0	—	0	—	0	2 “ $\frac{7}{8}''$ “	1.54
Cc	“	2.333	23800	“	$\frac{21}{9}$	54540	..	—	—	78340	6820	11.48	2 10″ chans.	11.46
Dd	“	1.333	13600	“	$\frac{15}{9}$	38960	..	—	—	52560	6220	8.45	2 8″ “	8.46
Ee	“	0.333	3400	“	$\frac{10}{9}$	25970	10	—	—	31970	5930	5.39	2 7″ “	5.40
Bb	“	0.5	5100	30940	1.0	30940	25	—	—	43780	10000	4.38	2 bars $3'' \times \frac{3}{4}''$	4.50
Chords.														
ac	22831	4.0	—	—	—	—	—	—	—	91320	10000	9.13	2 bars $5'' \times 1\frac{1}{8}''$ *	11.25
cd	“	7.0	—	—	—	—	—	—	—	159820	“	15.98	2 “ $5'' \times 1\frac{5}{8}''$	16.25
de	“	9.0	—	—	—	—	—	—	—	205480	“	20.55	4 “ $5'' \times 1\frac{1}{32}''$	20.62
ee	“	10.0	—	—	—	—	—	—	—	228310	“	22.83	4 “ $5'' \times 1\frac{5}{32}''$	23.12
BC	“	7.0	—	—	—	—	—	—	—	159820	7570	21.11	2 15″ chans.	24.00
CD	“	9.0	—	—	—	—	—	—	—	205480	7690	26.72	2 “ “	26.70
DE	“	10.0	—	—	—	—	—	—	—	228310	“	29.69	2 “ “	29.70
EE	“	10.0	—	—	—	—	—	—	—	228310	“	29.69	2 “ “	29.70
Top laterals														
BC	2550	3.0	—	—	—	—	—	7650	1.382	10570	15000	.70	1 rod 1″ diam.	.78
CD	“	2.0	—	—	—	—	—	5100	“	7050	“	.41	1 “ “	.78
DE	“	1.0	—	—	—	—	—	2550	“	3525	“	.24	1 “ “	.78
EE	“	0.0	—	—	—	—	—	—	—	0	—	0	1 “ “	.78
Bottom laterals														
ab	2550	4.0	10200	5100	$\frac{36}{9}$	20400		30600	1.458	44610	15000	2.97	1 rod $1\frac{3}{4}''$ sq.	3.06
bc	“	3.0	7650	“	$\frac{28}{9}$	15870		23520	“	34290	“	2.29	1 “ $1\frac{1}{2}''$ “	2.25
cd	“	2.0	5100	“	$\frac{21}{9}$	11900		17000	“	24790	“	1.65	1 “ $1\frac{5}{16}''$ sq.	1.72
de	“	1.0	2550	“	$\frac{15}{9}$	8500		11050	“	16110	“	1.07	1 “ $1\frac{1}{16}''$ “	1.13
ee	“	0.0	0	“	$\frac{10}{9}$	5670		5670	“	8270	“	0.55	1 “ $\frac{7}{8}''$ “	.77

* Gross.

Trigonometry and Navigation.

(See table of natural sines, cosines, tangents, and cotangents following.)

Employ that equation which contains the parts considered.

  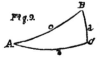

Solution of Plane Right Triangles (Fig. 7).

$$(1)\ \sin. A = \cos. B = \frac{a}{c}$$

$$(2)\ \cos. A = \sin. B = \frac{b}{c}$$

$$(3)\ \tan. A = \cot. B = \frac{a}{b}$$

$$(4)\ c^2 = a^2 + b^2$$

Ex. Given $A = 27° \ 30'$, and $b = 16.5$, to find a.
From (3), $a = b$ tan. $A = 16.5 \times .5206$ * $= 8.59$.
Ex. Given $a = 17.3$, and $c = 25.4$, to find B.
From (1), cos. $B = \dfrac{a}{c} = \dfrac{17.3}{25.4} = .6811$ † $= 47° \ 04'$, etc.

Solution of Plane Oblique Triangles (Fig. 8).

$$(5)\ \frac{a}{\sin. A} = \frac{b}{\sin. B} = \frac{c}{\sin. C}$$

$$(6)\ \tan. \tfrac{1}{2}(A - B) = \frac{(a - b)\ \tan. \tfrac{1}{2}(A + B)}{(a + b)}$$

$$(7)\ \tan. \tfrac{1}{2} A = \sqrt{\frac{(s - b)\ (s - c)}{s\ (s - a)}}$$

$$(8)\ s = \frac{a + b + c}{2}$$

* tan. 27° 30′ = .5095 + 37 × 3 = .5206.
† cos. 47° = .6820, (6820—6811) ÷ 2.2 = 4′.

Sin., cos., tan., or cotan. $A =$ sin., cos., tan., or cotan. (180°
$-A$).

Ex. Given $A = 50°\ 20'$, $B = 60°\ 10'$, and $a = 206.3$, to find b.
From (5), $b = \dfrac{a \cdot \sin. B}{\sin. A} = \dfrac{206.3 \times .8674}{.7698} = 232.51$.

Ex. Given $a = 90.2$, $b = 24.6$, and $C = 74°$, to find A and
B, $\tfrac{1}{2}(A+B) = \tfrac{1}{2}(180° - 74°) = 53°$, $(a+b) = 114.8$, $(a-b) = 65.6$.

From (6), tan. $\tfrac{1}{2}(A-B) = \dfrac{65.6 \times 1.3270}{114.8} = .7583$ ∴ $\tfrac{1}{2}(A-B)$
$= 37°\ 10'$, $A = 53° + 37°\ 10' = 90°\ 10'$, $B = 53° - 37°\ 10' = 15°\ 50'$.

Solution of Spherical Right Triangles (*Fig.* 9).

(9) $\sin. A = \dfrac{\sin. a}{\sin. c}$, $\sin. B = \dfrac{\sin. b}{\sin. c}$

(10) $\sin. A = \dfrac{\cos. B}{\cos. b}$, $\sin. B = \dfrac{\cos. A}{\cos. a}$

(11) $\cos. A = \dfrac{\tan. b}{\tan. c}$, $\cos. B = \dfrac{\tan. a}{\tan. c}$

(12) $\tan. A = \dfrac{\tan. a}{\sin. b}$, $\tan. B = \dfrac{\tan. b}{\sin. a}$

(13) $\cos. c = \cos. a\ \cos. b$

(14) $\cos. c = \cot. A\ \cot. B$

Ex. Given $c = 110°\ 50'$ and $A = 80°\ 10'$, to find a.
From (9), $\sin. a = \sin. c\ \sin. A = \sin. 69°\ 10' \times \sin. 80°\ 10' = .9346 \times .9853 = .9209$ ∴ $a = 67°\ 04'$ (Rule A).

Ex. Given $a = 116°$ and $b = 16°$, to find c.
From (13) $\cos. c = \cos. a\ \cos. b = \cos. 64° \times \cos. 16° = .4384 \times .9613 = .4214$ ∴ $c = 180° - 65°\ 05' = 114°\ 55'$ (Rule B).

Ex. Given $A = 60°\ 50'$ and $B = 57°\ 20'$, to find c.

From (14), cos. $c = \cot.\ 60°\ 50' \times \cot.\ 57°\ 20' = .5582 \times .6412$
$= .3578$ $\therefore c = 69°\ 02'$ (Rule B).

Rule A. An angle and its opposite side are always in the same quadrant, either both less or both greater than 90°.

Rule B. If a and b are in the same quadrant, c is less than 90°, and if in different quadrants, c is greater than 90°

Natural Sines, Cosines, Tangents, and Cotangents.

Deg.	Sine.	Diff. 10'	Cosine.	Diff. 10'	Tangent.	Diff. 10'	Cotangent.	Diff. 10'	Deg.
0	.0000	29	1.0000		.0000	29	∞		90
1	.0175	29	.9998	1	.0175	29	57.2900	4.775	89
2	.0349	29	.9994	1	.0349	29	28.6363	1.592	88
3	.0523	29	.9986	1	.0524	29	19.0811	7967	87
4	.0698	29	.9976	2	.0699	29	14.3007	4784	86
5	.0872	29	.9962	2	.0875	29	11.4301	3193	85
6	.1045	29	.9945	3	.1051	29	9.5144	2283	84
7	.1219	29	.9925	3	.1228	30	8.1443	1715	83
8	.1392	29	.9903	4	.1405	30	7.1154	1336	82
9	.1564	29	.9877	4	.1584	30	6.3138	1071	81
10	.1736	29	.9848	5	.1763	30	5.6713	878	80
11	.1908	29	.9816	5	.1944	30	5.1446	733	79
12	.2079	29	.9781	6	.2126	30	4.7046	622	78
13	.2250	29	.9744	6	.2309	31	4.3315	535	77
14	.2419	28	.9703	7	.2493	31	4.0108	465	76
15	.2588	28	.9659	7	.2679	31	3.7321	408	75
16	.2756	28	.9613	8	.2867	31	3.4874	361	74
17	.2924	28	.9563	8	.3057	32	3.2709	322	73
18	.3090	28	.9511	9	.3249	32	3.0777	289	72
19	.3256	28	.9455	9	.3443	32	2.9042	261	71
20	.3420	27	.9397	10	.3640	33	2.7475	237	70
21	.3584	27	.9336	10	.3839	33	2.6051	217	69
22	.3746	27	.9272	11	.4040	34	2.4751	199	68
23	.3907	27	.9205	11	.4245	34	2.3559	183	67
24	.4067	27	.9135	12	.4452	35	2.2460	169	66
25	.4226	26	.9063	12	.4663	35	2.1445	157	65
26	.4384	26	.8988	13	.4877	36	2.0503	146	64
27	.4540	26	.8910	13	.5095	36	1.9626	137	63
28	.4695	26	.8829	14	.5317	37	1.8807	128	62
29	.4848	26	.8746	14	.5543	38	1.8040	120	61
30	.5000	25	.8660	14	.5774	39	1.7321	113	60
31	.5150	25	.8572	15	.6009	39	1.6643	107	59
32	.5299	25	.8480	15	.6249	40	1.6003	101	58
33	.5446	24	.8387	16	.6494	41	1.5399	96	57
34	.5592	24	.8290	16	.6745	42	1.4826	91	56
35	.5736	24	.8192	16	.7002	43	1.4281	86	55
36	.5878	23	.8090	17	.7265	44	1.3764	82	54
37	.6018	23	.7986	17	.7536	45	1.3270	79	53
38	.6157	23	.7880	18	.7813	46	1.2799	75	52
39	.6293	23	.7771	18	.8098	48	1.2349	72	51
40	.6428	22	.7660	19	.8391	49	1.1918	69	50
41	.6561	22	.7547	19	.8693	50	1.1504	66	49
42	.6691	22	.7431	19	.9004	52	1.1106	64	48
43	.6820	21	.7314	20	.9325	54	1.0724	62	47
44	.6947	21	.7193	20	.9657	55	1.0355	59	46
45	.7071		.7071	20	1.0000	57	1.0000		45
Deg.	Cosine.	Diff. 10'	Sine.	Diff. 10'	Cotangent.	Diff. 10'	Tangent.	Diff. 10'	Deg.

Decimal Equivalents of English Measures.

Decimal of a Pound.						Decimal of a Cwt.				
s.	d.	£	s.	d.	£	qr. lbs.	cwt.	qr. lbs.	cwt.
	½	.0020	1	2½	.0604	1	.0089	1 21	.4375
	1	.0041		3	.0625	2	.0179	22	.4464
	1½	.0062		3½	.0646	3	.0268	23	.4554
	2	.0083		4	.0667	4	.0357	24	.4643
	2½	.0104		4½	.0688	5	.0446	25	.4732
	3	.0125		5	.0708	6	.0536	26	.4822
	3½	.0146		5½	.0729	7	.0625	27	.4911
	4	.0167		6	.075	8	.0714	2 0	.5
	4½	.0188		6½	.0771	9	.0803	1	.5089
	5	.0208		7	.0791	10	.0893	2	.5179
	5½	.0229		7½	.0812	11	.0982	3	.5268
	6	.025		8	.0833	12	.1071	4	.5357
	6½	.0271		8½	.0854	13	.1161	5	.5446
	7	.0291		9	.0875	14	.1250	6	.5536
	7½	.0312		9½	.0896	15	.1339	7	.5625
	8	.0333		10	.0916	16	.1429	8	.5714
	8½	.0354		10½	.0937	17	.1518	9	.5803
	9	.0375		11	.0958	18	.1607	10	.5893
	9½	.0396		11½	.0979	19	.1696	11	.5982
	10	.0416	2	0	.10	20	.1786	12	.6071
	10½	.0437	4	0	.20	21	.1875	13	.6161
	11	.0458	6	0	.30	22	.1964	14	.625
	11½	.0479	8	0	.40	23	.2054	15	.6339
1	0	.05	10	0	.50	24	.2143	16	.6429
	½	.0521	12	0	.60	25	.2232	17	.6518
	1	.0541	14	0	.70	26	.2322	18	.6607
	1½	.0562	16	0	.80	27	.2411	19	.6696
	2	.0583	18	0	.90	1 0	.2500	20	.6786
						1	.2589	21	.6875
						2	.2679	22	.6964
Decimal of a Shilling.						3	.2768	23	.7054
						4	.2857	24	.7143
						5	.2946	25	.7232
						6	.3036	26	.7322
d.	s.		d.	s.		7	.3125	27	.7411
½	.0417		6½	.5417		8	.3214	3 0	.75
1	.0833		7	.5833		9	.3303	1	.7589
1½	.125		7½	.625		10	.3393	2	.7679
2	.1667		8	.6667		11	.3482	3	.7768
2½	.2083		8½	.7083		12	.3571	4	.7857
3	.25		9	.75		13	.3661	5	.7946
3½	.2917		9½	.7917		14	.3750	6	.8036
4	.3333		10	.8333		15	.3839	7	.8125
4½	.375		10½	.875		16	.3929	8	.8214
5	.4167		11	.9167		17	.4018	9	.8303
5½	.4583		11½	.9583		18	.4107	10	.8393
6	.5					19	.4196	11	.8482
						20	.4286	12	.8571

Decimal Equivalents of English Measures (*Continued*).

Decimal of a Cwt. (*continued*).					Decimal of a Foot (*continued*).			
qr. lbs.	cwt.	qr. lbs.	cwt.		in.	ft.	in.	ft.
3 13	.8661	3 21	.9375		$7\ \frac{3}{8}$	.6146	$9\ \frac{3}{4}$	.8125
14	.8750	22	.9464		$\frac{1}{2}$	.6250	$\frac{7}{8}$	.8229
15	.8839	23	.9554		$\frac{5}{8}$	.6354	10	.8333
16	.8929	24	.9643		$\frac{3}{4}$	.6458	$\frac{1}{8}$	.8437
17	.9018	25	.9732		$\frac{7}{8}$	.6562	$\frac{1}{4}$	.8542
18	.9107	26	.9822		8	.6667	$\frac{3}{8}$	.8646
19	.9196	27	.9911		$\frac{1}{8}$	.6771	$\frac{1}{2}$	.8750
20	.9286	4 0	1.0		$\frac{1}{4}$	.6875	$\frac{5}{8}$	.8854
					$\frac{3}{8}$	.6979	$\frac{3}{4}$	.8958
					$\frac{1}{2}$	.7083	$\frac{7}{8}$	.9062
Decimal of a Foot.					$\frac{5}{8}$	.7187	11	.9167
					$\frac{3}{4}$	.7292	$\frac{1}{8}$	.9271
					$\frac{7}{8}$	.7396	$\frac{1}{4}$	.9375
in.	ft.	in.	ft.		9	.75	$\frac{3}{8}$	.9479
$\frac{1}{8}$	.0104	$3\ \frac{3}{4}$	.3125		$\frac{1}{8}$	.7604	$\frac{1}{2}$	.9583
$\frac{1}{4}$	.0208	$\frac{7}{8}$	.3229		$\frac{1}{4}$	.7708	$\frac{5}{8}$	.9687
$\frac{3}{8}$	.0312	4	.3333		$\frac{3}{8}$	.7812	$\frac{3}{4}$	.9792
$\frac{1}{2}$	.0417	$\frac{1}{8}$	.3437		$\frac{1}{2}$	.7917	$\frac{7}{8}$	.9896
$\frac{5}{8}$	.0521	$\frac{1}{4}$	.3542		$\frac{5}{8}$	.8021	12	1.0
$\frac{3}{4}$	.0625	$\frac{3}{8}$	.3646					
$\frac{7}{8}$	.0729	$\frac{1}{2}$	.3750					
1	.0833	$\frac{5}{8}$	.3854		**Decimal of a Lb.**			
$\frac{1}{8}$	.0937	$\frac{3}{4}$	.3958					
$\frac{1}{4}$	.1042	$\frac{7}{8}$	.4062					
$\frac{3}{8}$	.1146	5	.4167		oz.	lbs.	oz.	lbs.
$\frac{1}{2}$	.1250	$\frac{1}{8}$	.4271		$\frac{1}{2}$	.0312	$8\frac{1}{2}$	.5312
$\frac{5}{8}$	.1354	$\frac{1}{4}$	.4375		1	.0625	9	.5625
$\frac{3}{4}$	.1458	$\frac{3}{8}$	.4479		$1\frac{1}{2}$	.0937	$9\frac{1}{2}$	.5937
$\frac{7}{8}$	.1562	$\frac{1}{2}$	.4583		2	.1250	10	.625
2	.1667	$\frac{5}{8}$	.4687		$2\frac{1}{2}$	.1562	$10\frac{1}{2}$	.6562
$\frac{1}{8}$	.1771	$\frac{3}{4}$	.4792		3	.1875	11	.6875
$\frac{1}{4}$	.1875	$\frac{7}{8}$	.4896		$3\frac{1}{2}$	.2187	$11\frac{1}{2}$	.7187
$\frac{3}{8}$	.1979	6	.5		4	.25	12	.75
$\frac{1}{2}$	.2083	$\frac{1}{8}$	.5104		$4\frac{1}{2}$	.2812	$12\frac{1}{2}$	.7812
$\frac{5}{8}$	.2187	$\frac{1}{4}$	.5208		5	.3125	13	.8125
$\frac{3}{4}$	.2292	$\frac{3}{8}$	.5312		$5\frac{1}{2}$	.3437	$13\frac{1}{2}$	.8437
$\frac{7}{8}$	.2396	$\frac{1}{2}$	.5417		6	.375	14	.875
3	.25	$\frac{5}{8}$	.5521		$6\frac{1}{2}$	.4062	$14\frac{1}{2}$	.9062
$\frac{1}{8}$	.2604	$\frac{3}{4}$	.5625		7	.4375	15	.9375
$\frac{1}{4}$	.2708	$\frac{7}{8}$	.5729		$7\frac{1}{2}$	.4687	$15\frac{1}{2}$	.9687
$\frac{3}{8}$	.2812	7	.5833		8	.5		
$\frac{1}{2}$	.2917	$\frac{1}{8}$	.5937					
$\frac{5}{8}$	.3021	$\frac{1}{4}$	.6042					

THE K & E IMPROVED RECKONING MACHINE

A Perfect Mechanical Calculator.

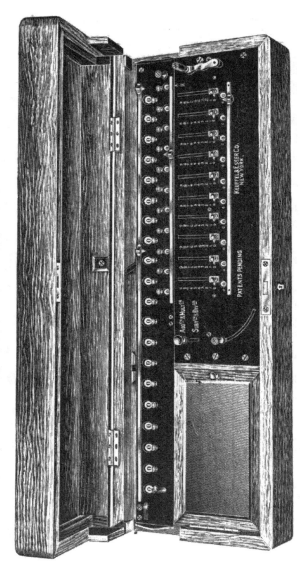

The Improved Reckoning Machine, which we now offer, represents the latest progress in the Manufacture of Merchanical Calculators. The new features include a device for simultaneously canceling the setting of all the keys in the key-plate, a row of openings below the key-plate grooves in which appear the settings of the several keys and four decimal point markers, arranged to slide on bars.

Within its limits this machine will solve practically any arithmetical problem with surprising rapidity and unfailing accuracy. The work of calculation is reduced to the simple process of setting the figure discs and shifters and turning the crank handle. Made in 3 sizes.

4005 N.	6 figures in lower, 7 in. middle and 12 in. upper row						$250.00
4006 N.	8	"	9	"	16	"	300.00
4007 N.	10	"	11	"	20	"	375.00

KEUFFEL & ESSER CO., New York.

www.ingramcontent.com/pod-product-compliance
Lightning Source LLC
LaVergne TN
LVHW012201040326
832903LV00003B/56